ENVIRONMENTAL SCIENCE, ENGINEERING AND TECHNOLOGY

RADIOACTIVE WASTE

SOURCES, MANAGEMENT AND HEALTH RISKS

Environmental Science, Engineering and Technology

Additional books in this series can be found on Nova's website under the Series tab.

Additional e-books in this series can be found on Nova's website under the e-book tab.

Environmental Science, Engineering and Technology

Radioactive Waste

Sources, Management and Health Risks

Susanna Fenton
Editor

New York

For permission to use material from this book please contact us:
Telephone 631-231-7269; Fax 631-231-8175
Web Site: http://www.novapublishers.com

Additional color graphics may be available in the e-book version of this book.

Library of Congress Cataloging-in-Publication Data

ISBN: 978-1-63321-731-7

Published by Nova Science Publishers, Inc. † New York

CONTENTS

PREFACE

This book discusses the perspectives of managing fusion radioactive materials. It also discusses the canister quandary, and the nuclear security system in Georgia

Chapter 1 - The question of "what to do with radioactive waste" has been raised frequently for both fission and fusion power plants. In recent years, fusion designers have become increasingly aware of the large amount of mildly radioactive materials that fusion generates. The search for a suitable solution has stimulated discussions about the origin and nature of fusion radioactive waste. Here, the authors discuss the problem rationally, support the argument with technical evidence, and propose a practical means to solve the problem. Lessons learned from other nuclear fields are valuable resources as the approaches and concepts for handling mildly active materials remain the same for fusion research, fission reactors, particle accelerators, and medical procedures.

Chapter 2 - Commercial spent nuclear fuel (SNF) is accumulating in the U.S. at the rate of approximately 2,000 metric tons (MT) per year, and a similar amount is being transferred each year from power plant fuel pools to dry storage. Most of the welded, sealed canisters used for dry storage can also be used for eventual transport to centralized storage or a geologic repository, hence these are called dual-purpose canisters (DPCs). By the year 2035 approximately half of all SNF in the U.S. is projected to be stored in DPCs. Unless these can be disposed of directly, they will all eventually need to be cut open and the SNF waste packaged in new containers.

DPC design has changed since they came into commercial use 20 years ago: they are larger and use various means to control criticality. Changes have occurred in parallel with advances in thermal and criticality analyses. Disposal has not been a factor in DPC design, due partly to the terms of contracts between utility companies and the government, which must ultimately disposition the SNF. A new canister design that is suitable for disposal also, has been addressed on two previous occasions: the multi-purpose canister (MPC) initiative of the 1990's, and the 2008 transport-aging-disposal (TAD) canister concept developed for a proposed repository in volcanic tuff. Both concepts were abandoned, hence SNF continues to accumulate in DPCs which are not purpose-designed or licensed for disposal.

The disposability of SNF canisters depends on effective safety strategies that ensure waste will be isolated in the repository, with sufficient heat dissipation and control of criticality. Such strategies will depend on disposal site characteristics, which may vary significantly for different host geologic media and disposal concepts. Over many thousands of years waste packages will cool, but will eventually be damaged by corrosion and may fill with ground water leading to the possibility of criticality. A framework for canister disposability that takes into account alternative safety strategies shows how repository programs in the U.S. and other countries have proceeded. It also shows what canisters options are available in the U.S. as alternatives to DPCs. Direct disposal of DPCs is also an option that is under study.

The potential advantages of canisters that are suitable for storage, transportation and disposal, or of direct disposal of existing DPCs, include simpler SNF management, lower cost, less secondary waste (e.g., used DPC hulls), and less worker exposure. Estimated cost savings in the U.S. depend on when such measures are introduced. Timely implementation is projected to save on the order of $10 billion in re-packaging costs. The capability to dispose of SNF in larger packages (e.g., capacity of at least 12 and up to 32 fuel assemblies from pressurized water reactors) would allow similar additional savings in disposal costs. The outlook for deploying multi-purpose canisters in the future depends on progress in disposal planning, and cooperation of the nuclear power utilities, the vendor industry, and government.

Chapter 3 - Georgia has received a difficult heritage from the former Soviet Union related to nuclear security situation. The country already took some steps to combat nuclear and radiation threats. Developing of nuclear forensics capability is one of the important steps to reach set goals to establish a nuclear security regime.

In: Radioactive Waste
Editor: Susanna Fenton
ISBN: 978-1-63321-731-7

Chapter 1

Perspectives of Managing Fusion Radioactive Materials: Technical Challenges, Environmental Impact and US Policy

Laila El-Guebaly[1*] and Lee Cadwallader[2]
[1]University of Wisconsin, Fusion Technology Institute, Madison, WI, US
[2]Idaho Falls, ID, US

Abstract

The question of "what to do with radioactive waste" has been raised frequently for both fission and fusion power plants. In recent years, fusion designers have become increasingly aware of the large amount of mildly radioactive materials that fusion generates. The search for a suitable solution has stimulated discussions about the origin and nature of fusion radioactive waste. Here, we discuss the problem rationally, support the argument with technical evidence, and propose a practical means to solve the problem. Lessons learned from other nuclear fields are valuable resources as the approaches and concepts for handling mildly active materials remain the same for fusion research, fission reactors, particle accelerators, and medical procedures.

[*] Corresponding author: elguebaly@engr.wisc.edu.

INTRODUCTION

Fusion has long been envisioned as possessing an inherent advantage for benign environmental impact, mainly due to the absence of high-level waste (HLW) generation. However, fusion tends to generate a sizable amount of mildly radioactive materials. Such a potential problem has been overlooked in early fusion studies and/or relegated to the back-end as only a disposal issue in low-level waste (LLW) repositories, adopting the preferred radioactive waste (radwaste) management approach of the 1960s. The large volume of fusion LLW generated during operation and after decommissioning will fill existing US LLW repositories rapidly. This means that shallow land burial is not a viable option for fusion and it is essential to think of a more environmentally attractive framework to keep the fusion LLW volume to a minimum. This is important to the future of fusion energy. Concerns about the environment, radwaste burden for future generations, lack of geological repositories, and high disposal cost directed our attention to recycling of the radioactive materials (for reuse within the nuclear industry) and clearance (the unconditional release to the commercial market if materials contain traces of radioactivity).

In the past, the recycling and clearance options have been investigated by fusion researchers in the late 1980s and 1990s, focusing on selected materials or components [1,2,3], and then examining almost all fusion components in the late 1990s and 2000s [4-8]. In recent years, the recycling and clearance approaches became more technically feasible with the development of advanced radiation-hardened remote handling (RH) tools that can recycle highly irradiated materials [9,10,11] along with the introduction of the clearance category for slightly radioactive materials by national and international nuclear agencies [12,13]. Such recent advances encouraged many fusion designers in the US and throughout the world [14-21] to apply the recycling and clearance approaches to all fusion components that are subject to extreme radiation levels: very high level near the plasma and very low level at the biological shield (bioshield). If integrated properly at an early stage of the design process, fusion will eventually reach the ultimate goal of radwaste-free energy source. At present, the US experience with recycling/clearance is limited, but will be augmented significantly by advances in fission reactor dismantling, spent fuel reprocessing, and bioshield clearance before fusion is committed to commercialization in the second half of the 21st century.

Beginning in 2000, numerous fusion studies indicated the recycling/clearance approach is relatively easy to envision and apply from a science perspective. To support this argument, we applied all three scenarios (disposal, recycling, and clearance) to the most recent US magnetic fusion power plant: ARIES-ACT-2 [22] – a tokamak with a net electric power of 1000 MW_e, 9.75 m major radius, 2.44 m minor radius, and fusion power of 2637.5 MW. Ferritic steel (FS) is the main structure for all in-vessel components, except the divertor, which is made of a combination of W alloy and FS. An isometric view of ARIES-ACT-2 is shown in Figure 1 while the main components comprising the radial and vertical builds are identified in Figure 2. The LiPb/He dual-cooled blanket breeds sufficient tritium for plasma operation, recovers the neutron energy, and protects the shield for the entire plant life (40 full power years). The blanket and shield help protect the manifolds and vacuum vessel (VV) and all four components protect the superconducting magnets for life. The machine average neutron wall loading (NWL) is 1.5 MW/m^2. Based on the 2.2 MW/m^2 peak NWL at the outboard (OB) midplane, the FW, blanket, and divertor will be replaced every 8.65 full power years (FPY) [22] while the outer components (outer segment of OB blanket, structural ring (SR), VV, low-temperature (LT) shield, magnet, cryostat, and bioshield) are expected to operate successfully for the entire plant life (40 FPY) with an overall system availability of 85%.

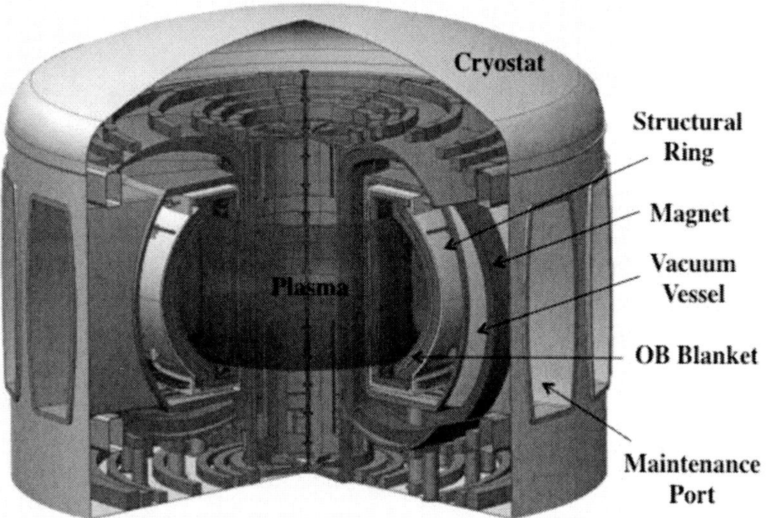

Figure 1. ARIES-ACT-2 isometric view.

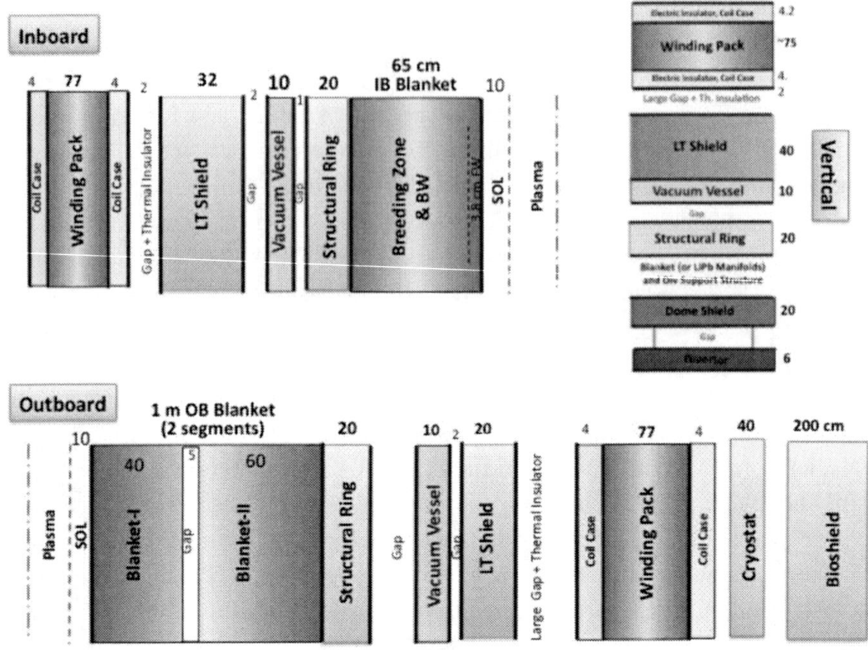

Figure 2. ARIES-ACT-2 radial build at midplane and vertical build through divertor dome.

Here, we focus on the technical challenges of managing fusion radioactive materials and the projected US policy governing the future of environmentally attractive nuclear facilities. Since the proposed approaches and concepts remain the same for fusion research as well as for fission reactors, particle accelerators, medical procedures, or other nuclear areas, we also cover the fission experience, Nuclear Regulatory Commission (NRC) and Department of Energy (DOE) rules, NRC waste classifications, and licensed LLW and HLW repositories in the US and abroad.

2. Concerns about Fusion Radioactive Inventory

Fusion reactions within the plasma are dissociated from the surrounding wall materials. To hold the fusion promise of electricity production with low environmental impact, designers had the freedom to select low-activation materials for all components comprising the fusion power core. Since the early

1970s, materials research has been conducted to find the optimal structure for fusion power plant components while satisfying several design criteria, including fabrication, structural integrity under irradiation, and activation requirements. Historically, the early low-activation criteria were based on releases during accidents and low-level waste disposal for short-lived materials that require only near-surface burial and relatively short regulatory oversight of ~100 years. Attempts were made in recent years to incorporate new improvements that deal with recycling and clearance to minimize the volume of radwaste assigned to geologic repositories. Such requirements limited the technology choices for fusion and motivated researchers to develop four main classes of low-activation structural materials [23,24]: ferritic/martensitic and oxide dispersion strengthened (ODS) steels, silicon carbide/silicon carbide (SiC/SiC) composites, tungsten alloys, and vanadium alloys. The reduced activation ferritic/martensitic (RAFM) steel is the most mature with extensive experience in steelmaking at the national and international levels. The vanadium alloy represents an option that is only attractive with the self-cooled lithium blanket. W-TiC and W-La_2O_3 are the prime W alloy candidates for divertor applications [25]. More advanced W alloys (such as fiber-reinforced W/W composites and nano-structured W) are still under development. The least developed SiC/SiC composites offer higher operating temperatures (to enhance the economics) and very low decay heat – an outstanding safety advantage. Upon exposure to fusion neutrons, the four candidate structural materials become radioactive at different degrees depending on their activation and decay characteristics. Figure 3 shows the specific activity of the four structural materials subject to the same radiation environment at the first wall (FW) of a typical tokamak. A realistic level of impurities was included in the compositions of all materials. Within one day, the SiC/SiC activity drops significantly while the activities of all materials decrease by 6-7 orders of magnitude in 100 years after plant shutdown – a salient safety feature of low-activation materials.

There is a potential concern regarding the radioactive inventory that fusion generates. To put matters into perspective, we compared in Figure 4 the power core volumes of the ITER [26] experimental device, the advanced ARIES power plants (ARIES-ACT-1&2) [22,27] and the European Power Plant Conceptual Study (PPCS) [28] to ESBWR (Economic Simplified Boiling Water Reactor) – a Gen-III$^+$ advanced fission reactor [29]. As noted, fusion power cores generate sizable volumes of mildly activated materials compared to fission. In recent years, fusion designers paid much attention to the waste management issue associated with the large volume of radioactive materials discharged from fusion power plants. This has been accomplished by efforts in the US and throughout the world. Essential steps included reshaping the fusion waste management approach and maximizing the reuse of materials through recycling and clearance in order to minimize the environmental impact, free ample space in repositories and, in the long run, save fusion millions of dollars for the high disposal cost.

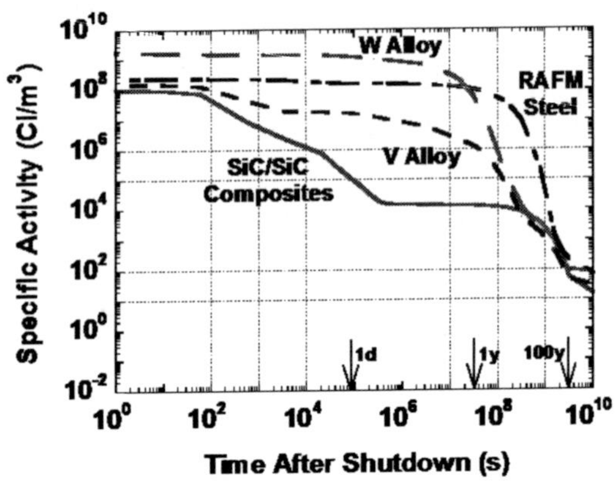

Figure 3. Specific activity of FW made of RAFM steel, SiC/SiC composites, V alloy, or W alloy.

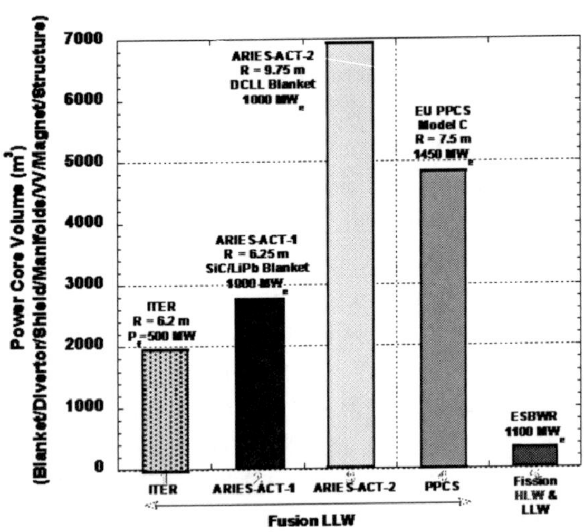

Figure 4. Comparison of radioactive waste from power core of fusion and fission designs (actual volumes of components; not compacted; no replacement).

Surrounding the fusion power core is the bioshield – a 2 m thick, steel-reinforced concrete building constructed to withstand natural phenomena and essentially protects the public and workers against radiation. Being away from the neutron source, the bioshield is subject to low radiation and contains very low radioactivity. However, its volume (not shown in Figure 4) dominates the waste stream. Since burying such a huge volume of slightly activated materials in geologic repositories is impractical, regulatory agencies around the globe suggested the clearance concept where such components could temporarily be stored for the radioactivity to decay, then released to the commercial market for reuse as shielding blocks, concrete rubble base for roads, deep concrete foundations, non-water supply dams for flood control, etc.

3. NRC and DOE Standards and Regulations

The US Nuclear Regulatory Commission (NRC) is responsible for radioactive waste generated by commercial production of electricity and other non-military uses of nuclear materials. The NRC has established the requirements for fission power plants and other radioactive waste management and storage in the US. Since fusion radioactive waste is essentially the same as the wastes generated by fission power plants, these requirements are applied to fusion as well. The Code of Federal Regulations (CFR) gives regulations on radioactive waste. 10CFR60 [30] discusses disposal of high-level radioactive waste in deep underground repositories. 10CFR61 [31] discusses licensing of facilities for land disposal of radioactive waste. 10CFR71 [32] gives the requirements for packaging, preparation for shipment, and transportation of radioactive materials. Other US government agencies, the DOE, Environmental Protection Agency, the Department of Transportation, and the Department of Health and Human Services, also have roles in regulating radioactive waste. The DOE has the responsibility of dealing with radioactive waste from the creation of nuclear weapons and the cleanup of the facilities that manufactured these weapons. The DOE also deals with the waste from materials testing reactors and particle accelerators. Typically, the DOE operations and cleanup have higher volume waste streams than those from commercial power plant operation and decommissioning.

The US NRC has given some attention to the issue of reducing the amount of radioactive waste generated when decommissioning nuclear power plants. Regulatory Guide 1.86 [33] discusses the release of material to commercial industry. A goal of decommissioning nuclear facilities is to minimize waste volumes and recycle as much of the materials as practical [34].

The DOE has a qualitative guide [35] on releasing very mildly radioactive materials, particularly metals, to return them to commercial industry as scrap metal. This is desirable for several reasons – to reclaim use of metal resources, to reduce the volume of LLW, and generate income for government activities. Scrap metal is a valuable commodity since it has already been mined, purified, and alloyed. However, there are strict rules on allowing very mildly radioactive material into the commercial market. Turner [36] discussed the scrap metal industry concerns with radioactive materials. The industry is concerned about radioactive contamination of its facilities and exposure to its personnel, who are not radiation workers, the cost of decontamination, and also the loss of consumer confidence in the safety of steel, copper, aluminum, lead, and other metal consumer products that might become even mildly contaminated. There have been some events of radioactive sources being smelted into scrap metal [37,38]. In 2000, then Secretary of Energy Mr. Bill Richardson imposed an agency-wide suspension on unrestricted release of scrap metal originating from radioactive areas at DOE facilities for the purpose of recycling. This suspension followed the Oak Ridge National Laboratory giving a contract to British Nuclear Fuels Limited to decommission the K-25 gaseous diffusion plant with its very large quantities of mildly contaminated steel, aluminum, copper, and nickel. The contractor was to underwrite part of the decommissioning cost with the sale of uncontaminated scrap metal. There were concerns from the public and the scrap metal industry about the DOE process for segregation of uncontaminated metals and the potential effects of radioactivity entering into the scrap metal streams. The DOE was tasked to develop and implement improvements in its release process. In 2012, the DOE released a programmatic environmental assessment on recycling mildly contaminated metal [39]. A new DOE directive was also released in 2011 that defines a 1 mSv (100 mrem) total effective dose to members of the public from DOE radiological activities, including exposure to residual radioactive material subsequent to any remedial action or clearance of DOE property [40]. The DOE directive does not address volumetrically activated materials, just surface contaminated materials. It is noted that public comments have indicated disbelief in the calculations used to assess potential doses [41]. As of this writing, the DOE has not yet resumed release of scrap metal.

4. Disposal in Geological Repositories

The disposal option attracted growing attention in the late 1960s as the preferred option for handling radioactive waste. To date, and after many years in the energy market, the nuclear industry continues to struggle with the management of radwaste as the prediction of geological and climatology conditions is less accurate for longer times into the future.

4.1. NRC Classification of Radwaste

There are several types of radioactive waste defined by the US NRC. HLW is defined as spent nuclear fission fuel and any reprocessing liquids or residues. HLW is typically very radioactive and can have high heat generation. HLW requires robust shielding (i.e., such as deep geologic storage) to provide for safety. Transuranic (TRU) waste is composed of elements above uranium in the periodic table of elements. It tends to be much less radioactive than HLW, with lower internal heat generation and low particle emission, but the TRU radionuclides generally have very long half-lives. Another category of waste is uranium mill tailings, which are typically very low radioactivity and almost no heat generation. There is also low-level waste (LLW), which is basically any waste that does not fit into the other listed categories. Generally, LLW is not highly radioactive, has very low heat generation, and can be safely disposed of in a near-surface repository. LLW has three classes, called A, B and C [42]. Class A waste is the least radioactive, and is typically composed of biomedical waste from hospitals, waste from power plants, and waste from industries using radioactive materials. The Class A power plant waste tends to be used anti-contamination clothing (coveralls, booties, hoods, gloves), low-contaminated resins from coolant cleaning systems, and general trash that could be radioactively contaminated. Class B waste is higher specific activity, that is more radioactivity per gram, than Class A waste and can be composed of filters and filter residues, concentrated sludge from evaporated coolant, and other refuse. Class C waste is the most radioactive class of LLW. It often consists of contaminated tools or broken parts, used ion exchange resins, spent sealed sources, used coolant seals, and other wastes [43]. There is another type of LLW, referred to as greater than Class C (GTCC), which is low level waste that does not meet the requirements of Class C waste.

4.2. Existing Commercial Repositories in US and Abroad

At present, LLW is disposed by packaging the waste and burial in shallow trenches; this is referred to as near-surface disposal or shallow land burial. Trenches or excavations are typically a few meters deep and are covered by a protective cap of some form of membrane and stones or soil to keep water out of the trench. There are four licensed low level waste disposal facilities in the US. These are: EnergySolutions Barnwell Operations facility at Barnwell, South Carolina (LLW Classes A to C), the US Ecology site at Richland, Washington (LLW Classes A to C), EnergySolutions Clive Operations facility at Clive, Utah (LLW Class A only), and the Waste Control Specialists (WCS) facility at Andrews, Texas (LLW Classes A to C). It is noted that since 2008 the Barnwell facility will only accept waste from just a few states that are part of a compact agreement, Connecticut, New Jersey, and South Carolina. The Richland facility accepts waste from these states: Alaska, Colorado, Hawaii, Idaho, Montana, New Mexico, Nevada, Oregon, Utah, Washington, and Wyoming. The Clive facility accepts waste from all regions of the US. The Andrews facility accepts waste from Texas compact states (Texas and Vermont) and other waste generators with permission from the compact agreement states. Typically, about 2 million cubic feet (~56,700 m^3) of LLW is disposed of annually at these sites, with more than 95% of that waste going to the Clive Class A LLW facility. Figure 5 depicts the sites of the four large-scale commercial LLW repositories in the US. There is some indication that three out of the four US commercial LLW repositories will reach their maximum capacity and may close by mid-century before commissioning the first fusion power plant.

The US had been politically unable to open a deep geologic HLW repository for commercial spent nuclear fuel (SNF), despite the work at the Yucca Mountain Project in Nevada. Currently, the US has one operational deep geologic repository, the Waste Isolation Pilot Plant (WIPP) near Carlsbad, New Mexico. This demonstration facility is operated by the DOE and serves as a permanent repository for defense-generated TRU waste. The reader will note that the TRU waste at WIPP is low heat generation so the WIPP is not truly a HLW repository. The facility opened in March 1999 and by the end of 2013 had emplaced 8.52E+04 m^3 of TRU waste from several DOE sites in the US [44]. It is noted that after an uneventful 15 years of operation, WIPP had two events in February 2014. In one event, small amounts of radioactivity were released to the atmosphere, but the maximum estimated doses to workers and public were well below the allowable limits

[45]. The WIPP facility is temporarily closed to allow decontamination and system modifications. Presently in the US, spent nuclear fuel is stored at the nuclear power plants that used the fuel. Newer spent fuel that has high heat generation is stored in water pools to allow heat removal. Older spent fuel that has had up to 30 to 40 years to decay has lower heat generation and is stored in dry casks at ground level. This is referred to as interim storage, while a deep geologic repository is viewed as permanent storage. The first dry cask storage license in the US was granted in 1986. As of 2013, there are 54 operating, general licensed Independent Spent Fuel Storage Installations (ISFSIs) of dry casks at fission power plant sites in the US, and 15 more operating, specific licensed ISFSIs located at or away from power plant sites.

Most countries have not been politically able to open deep geologic repositories and are using pools and cask storage for SNF. Some countries have operated underground research or test facilities for waste disposal, including boreholes, former underground mines, and other test cases [46]. Some countries define a class of waste as intermediate level waste (ILW), where the radioactivity is reduced by half in 30 years. Some countries have operated deep geologic repositories for LLW and ILW. Most of these national repositories are large, with storage capacities in the 0.5 to 4.5 million cubic meters range:

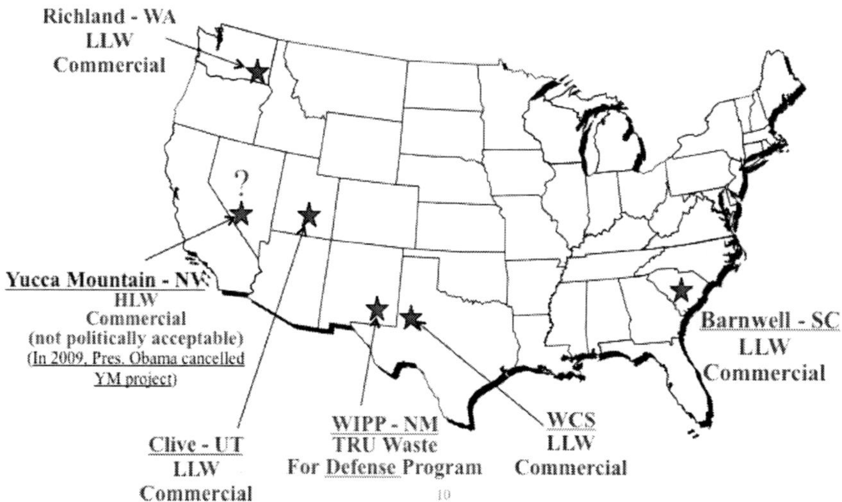

Figure 5. Sites for large-scale repositories in the US.

- **Germany** operated the Endlager für radioaktive Abfälle Morsleben (ERAM) facility in an abandoned salt mine from 1981-1998, placing 36,800 m^3 of waste: about 28,500 m^3 was solid waste and 8,300 m^3 was liquid waste [47]. Germany also has a geologic repository for cask storage of low-heat vitrified waste from reprocessing of their spent fuel (reprocessed at La Hague, France) at Gorleben, and a dry cask spent fuel interim storage facility at Ahaus. The former Konrad iron mine near Salzgitter, Germany is planned to be used to store LLW and ILW waste that does not generate heat.
- **United Kingdom** operates a LLW repository at Drigg in Cumbria.
- **Spain** has the El Cabril LLW and ILW facility in the foothills of the Sierra Albarrana.
- **France** has the Centre de l'Aube near Troyes.
- **Finland** has LLW disposal adjacent to the Olkiluoto and Loviisa nuclear power plants.
- **Sweden** operates the Swedish Final Repository (SFR) for LLW and ILW; it is 50 to 120 m depth. The SFR has 63,000 m^3 volume capacity and has operated since 1983. Sweden also has a pool storage facility for spent fuel at Oskarshamn [48].
- **Switzerland** has an interim storage facility for HLW (stored in casks in shafts), ILW and LLW (near surface storage) that has been open since 2001. The facility is operated by the company Zwischenlager Wuerenlingen AG (ZWILAG), and it is adjacent to the Paul Scherrer Institute near the town of Wuerenlingen. HLW is stored in dry casks for 30 to 40 years. ZWILAG also has a volume-reduction plasma plant for thermal decomposition of the low level waste (see www.zwilag.ch).
- **Japan** has an LLW disposal center at Rokkasho in northern Japan.
- **S. Korea** finished construction of the Gyeongju waste disposal facility in 2013, and it is undergoing commissioning in 2014. This is an 80-m underground silo and cavern-based facility for low and intermediate level waste disposal that supports the country's 21 fission power plants. Initial storage volume is 100,000 drums and expansion to 800,000 drums is possible. Each power plant stores its 200-liter waste drums on-site until these can be shipped to Gyeongju, which is located southeast of Seoul. Used nuclear fuel is also stored at power plant sites; in pools for the first 6 years and then transferred to dry cask storage. There is a plan to open a used nuclear fuel interim storage facility by 2024 (see www.world-nuclear.org).

There are 43 countries in the world that must dispose of spent nuclear fuel as HLW. Twenty-five of these countries have decided that burying the HLW deep underground is the safest and most secure solution. No country at present has an operational deep geologic repository for HLW. Like the US, nearly all countries are currently using interim above-ground storage of HLW until suitable facilities are opened. Several countries are planning, characterizing, or developing their own sites for deep geologic repositories [49]. There are several choices for the type of geologic repository, including salt, crystalline rock, clay, and volcanic tuff [50]. Some other options discussed for long-lived radioactive waste disposal are: borehole disposal (drilling a deep hole, placing a cask, backfilling the hole), disposal by melting into rock at some depth (e.g., 2 km depth), sub-seabed disposal, disposal in ice sheets, disposal in outer space, and disposal in subduction zones (where one tectonic plate dives beneath the other). Some of these options are not possible due to international treaties. For example, launching waste into outer space and using Antarctica for a disposal site are precluded by treaties [49]. The option that has had the most study is geologic disposal (e.g., up to 1 km depth) using casks.

4.3. Present Practices of Decommissioning Fission Waste Streams

There are three types of waste streams to address: solid, liquid, and gas. As mentioned in Section 4.2, low-level fission reactor solid waste, LLW, is disposed of in near-surface burial sites such as the Clive, Utah facility. This is true for operational waste from power plants as well as the decommissioning waste. Liquid waste can be processed for volume reduction (e.g., filtering, reverse osmosis, evaporating the water from the liquid) and the liquid can be immobilized on absorbent material such as vermiculite, then packaged for burial as LLW. Gaseous waste streams tend to be held in tanks or pipes at power plants, giving a delay time for radionuclides, such as Argon-41 to decay, then the greatly reduced radioactive or stable nuclide gases are released to the atmosphere.

Decommissioning waste is a special case due to the large volume and the variety of types of waste. When a nuclear facility reaches its end of life, it is dismantled into its component parts. There are several waste streams, including concrete, metal, and water processing wastes (filters, sludge) [51]. The concrete is often only surface contaminated. Contamination is referred to as either smearable or fixed. Smearable means that the radionuclides can be removed with relative ease, such as by washing with water or by chemical decontamination. Fixed contamination means the contamination adheres and must be removed by removing the top layer of the concrete, usually by mechanical means such as scrabbling. The concrete can be reduced to rubble by mechanical means or by demolition. If clearable, concrete from decommissioning can be used in construction of new roads, concrete foundations, etc.

Waste materials are either cleaned sufficiently for clearance (free release), or they are recycled or buried. The disposition of waste is governed by the level or radioactivity compared to the release standards, the demand for and acceptance of recycled materials, and the cost of burial [52]. Currently, the zero tolerance policy for even mildly activated or contaminated materials in the metal industry means that waste materials are segregated into non-radioactive and radioactive streams. Smith [53] stated that the Plum Brook reactor in Ohio was decommissioned and all of the concrete stayed on site as backfill into the void of the reactor. An estimate of 95% of all demolished material by weight and volume were reused or recycled, including all of the demolished concrete and metals. All scrap steel was scanned for radiation before being sent to scrap metal yards. Contaminated material was placed in boxes for disposal at the Clive facility in Utah.

Banovac [34] points out that each fission reactor decommissioning project is unique due to the plant, the location, state laws, and other factors. A few examples follow. At the Trojan plant in Oregon, all concrete structures were decontaminated and released for unrestricted use. The Trojan decontamination and decommissioning (D&D) activity only disposed of 12,375 m^3 of LLW due to its minimization of waste volumes and recycling. The Big Rock Point D&D staff determined that about half of the concrete was non-impacted (never had the potential for neutron activation or exposure to licensed radioactive material) so it was reused. The other half of the demolition debris, 19.16 Mkg of predominantly concrete and some metals was mildly contaminated or activated and the licensee requested disposal in a State of Michigan Type II landfill. Since the calculated potential annual dose to a worker or member of the public at or near the landfill was less than 0.01 mSv (1 mrem) the NRC granted permission. At the Rancho Seco plant in California, the licensee chose to defer disposal of Class B and C waste, making an Interim Onsite Storage Building. The licensee also stored its GTCC waste at their ISFSI. For the Maine Yankee plant to achieve site release for unrestricted use, they removed all above-ground structures to a depth of 0.91 m below grade. About 100,000 m^3 of radioactive waste was disposed of offsite.

4.4. Problems with the Disposal Approach

There are a number of problems in disposal of radioactive materials from plant operation as well as D&D:

- There is no well-developed D&D disposal approach since each power plant is unique in its size, location, local laws, and other factors. As more aged power plants undergo D&D, perhaps more commonalities will emerge and the approach will become more standardized.
- US government attempts to make regulations on debris free release or clearance for unrestricted use have not been successful due to public concerns over radiation exposure (see Section 5.2.3), so valuable materials such as steel and other scrap metals cannot be recycled or cleared. The need to dispose of these metals as LLW or even in a landfill creates another cost of decommissioning.
- The existing government rules are sometimes inconsistent. Regulatory Guide 1.86 [33] is viewed as an official document giving release guidance, and its allowable contamination for radioisotopes such as Co-60 is 5,000 dpm/100 cm^2, while a more recent NRC document [12] gives a much more restrictive value of 280 dpm/100 cm^2 for Co-60 [54].
- The limited waste repository space and difficulty of building new ones. Most countries are having political difficulty in opening deep geologic repositories for spent fission fuel and other HLW; countries are using ISFSI's. There is also resistance to opening near-surface LLW repositories. In the US, over 56,000 m^3 of LLW are disposed of annually. The disposal sites fill over time and finding a location whose residents are amenable to having a facility, plus the high cost of licensing and constructing new facilities, is difficult.
- To scan D&D debris for its releasability, using radiation detectors to search for activity on the order of 10 pCi/gram or less in irregularly shaped debris or components in quantities of thousands of kg at a time is a challenge.
- The US deep geologic repository, WIPP, had two recent events in February 2014 that undermine confidence in underground storage: 1) a fire in a haul truck created a need to evacuate personnel, and 2) a waste drum breach that contaminated a small portion of the mine. In the first event, a diesel powered salt haul truck caught fire, perhaps from hydraulic fluid or diesel fuel contacting hot surfaces of the

engine exhaust system. Despite attempts to extinguish the fire, it spread to the front tires of the truck, filling the underground with smoke and soot. There were 87 workers in the mine at the time and all evacuated; some required treatment for smoke inhalation. No radiation was released in this event [55]. In the second event, there was a radiological release that was apparently due to a thermal excursion in a waste drum. The mine ventilation system swept the released Pu-239, Pu-240, and Am-241 material to the surface where the ventilation air was routed to high efficiency filters. Small amounts of these alpha and beta emitters were released to the atmosphere by small leaks past the filters. The workers closest to the release have been estimated to have received 0.1 mSv (10 mrem), and the maximum estimated public dose is less than 0.01 mSv (1 mrem). These doses are well below allowable limits [45]. The WIPP facility is temporarily closed for a year or perhaps longer to decontaminate the tunnel and replace the ventilation system. In general, a fire in a waste storage facility is always a safety concern since fire heat can overpressurize waste drums, waste drum breach and radiological release could result in the ventilation air sweeping radionuclides back to the atmosphere, negating the advantage of > 600 m depth underground.

4.5. Relative Estimate of Disposal Cost

The US NRC has published data on costs for waste disposal [56]. As an example, the Class A LLW facility at Clive responded with some costs:

Large components	\$350 per cubic foot
Debris	\$145 per cubic foot
Oversize debris	\$165 per cubic foot
Resins/filters	\$460 per cubic foot
Combustibles	\$575 per cubic foot
Evaporator bottoms	\$ 14 per cubic foot.

These are disposition costs that do not include shipping. These costs do not include taxes (roughly estimated to be 10%), and do not have any volume discounts, which could lead to savings. The costs rose about 15.5% from 2010 to 2012.

LLW has generally ranged between $100 and $1,000 per cubic foot for disposal [54]. The US NRC has guidance on estimating disposal costs during D&D [57,58] and the US Code of Federal Regulations has rules on the minimum amount of funding to have on hand for decommissioning a fission power plant. 10CFR50.75 gives the minimum funding a licensee must have in its decommissioning fund as $105M for a large pressurized water reactor and $135M for a large boiling water reactor (both values are in 1986 dollars). Accounting for inflation, these values are about $224M and $288M, respectively. Presently, fission reactor decommissioning generally costs in the range of $300M to $400M for a plant. The licensee must make up the difference. Figure 6 displays a schematic of the essential underground layers for disposal of LLW and HLW that partially reflects the high disposal cost. An important lesson learned from the fission experience is that fusion should avoid generating HLW (since there is design latitude in materials selection), minimize the Class C LLW, and tolerate any Class A LLW.

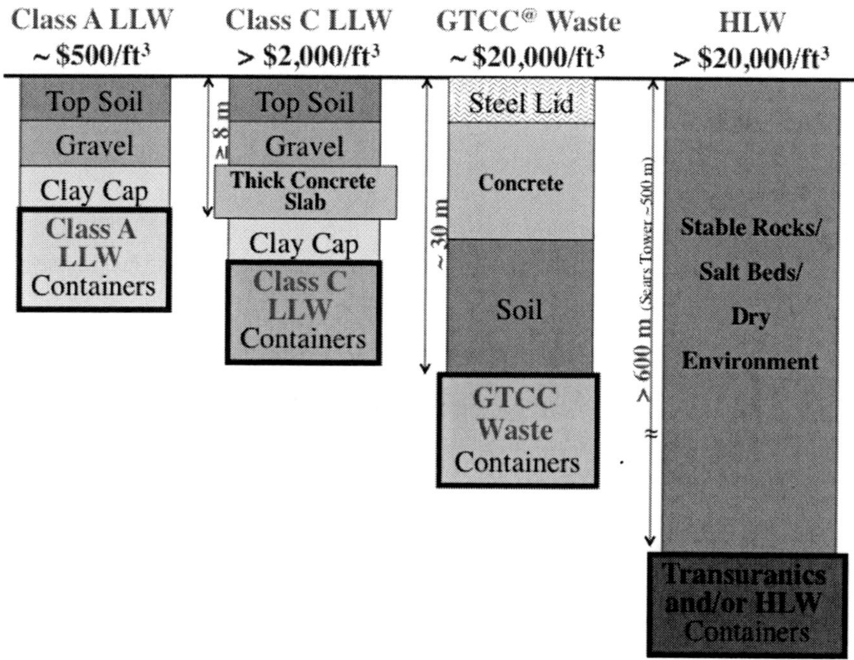

Figure 6. Schematic of below ground layers for disposal of LLW and HLW. Estimated cost partially covers preparation, characterization, packaging, interim storage, transportation, licensing, disposal and monitoring.

4.6. Technical Aspects of Disposal, Critical Issues, and Needs

For fusion designs, the waste disposal rating (WDR) has been evaluated for fully compacted waste using the waste disposal limits developed by NRC 10CFR61 [42] and Fetter [59]. The NRC waste classification is based largely on radionuclides that are produced in fission reactors, hospitals, research laboratories, and food irradiation facilities. In the early 1990s, Fetter and others [59] performed analyses to determine the Class C specific activity limits for all long-lived radionuclides of interest to fusion using a methodology similar to that of 10CFR61 [42]. Although Fetter's calculations carry no regulatory acceptance, they are useful because they include fusion-specific isotopes. All fusion components should meet both NRC and Fetter's limits until the NRC develops official guidelines for fusion waste. Also, the WDR is evaluated at 100 y after shutdown, allowing the short-lived radionuclides to decay. A WDR < 1 means LLW and WDR > 1 means HLW.

All ARIES-ACT-2 components qualify as Class C LLW. The LT shield and magnets are less radioactive than the in-vessel components and qualify as Class A LLW – the least hazardous type of waste. Excluding the clearable components (cryostat and bioshield), ~40% of the ARIES-ACT-2 waste (blanket, SR, VV, and divertor) qualifies as Class C LLW. The dominant radioisotope at 100 y after shutdown is ^{94}Nb (from 0.5 wppm Nb impurity in FS). The remaining ~60% (LT shield and magnet) would fall under the Class A LLW category. Several critical issues for the disposal option can be identified based on the assessment of disposal situation in the US. These are listed below along with specific needs for the fusion community:

Disposal issues:

- High disposal cost (for preparation, characterization, packaging, interim storage, transportation, licensing, and disposal)
- No HLW repositories
- Limited capacity of existing LLW repositories
- Political difficulty of building new repositories
- Prediction of repository's conditions for long time into future
- Radwaste burden for future generations.

Disposal needs:

- Revised activity limits for fusion radioisotopes issued by NRC
- Repositories designed for T-containing materials

- Reversible disposal process and retrievable waste (to gain public acceptance and ease licensing).

5. Alternate Approaches to Disposal

The recycling and clearance approaches could significantly minimize the volume of radioactive materials assigned to geologic disposal. In order to provide a broader perspective of the relevant issues involved in both processes, we identified at the end of Sections 5.1 and 5.2 the key issues and needs for recycling and clearance, respectively. As a step forward, a dedicated research and development (R&D) program should tackle these issues to help optimize the radwaste management scheme further.

5.1. Recycling

Recycling is a real indicator of whether any nuclear industry is serious about reducing its radioactive waste volume. The pros and cons debates will continue for sometime into the future as some arguing recycling could result in substantial technological difficulties, while others claiming the environmental benefits far outweigh any adverse effects. Hrncir [60] stated that recycling mildly contaminated metals has the environmental benefits of saving the metal itself – reclaiming its intrinsic value, and protecting the environment from the negative effects of ore mining and ore processing. Regarding the economic aspect of recycling, a Russian study concluded that recycling is cheaper than disposal [61]. In the US, there was a cost saving in recycling lead shielding bricks at the Idaho National Laboratory (INL) versus disposal in LLW repositories, as will be discussed shortly.

The recycling process includes storing in continuously monitored facilities, segregation of various materials, crushing, melting, and re-fabrication [16]. The highly radioactive components require special shielding during handling and transportation. Some may even need cooling for several days to remove the decay heat. Note that most fusion radioactive materials contain tritium that could introduce serious complications to the recycling process. Detritiation treatment prior to recycling is necessary for all fusion components.

Many nuclear nations are moving toward recycling. At present, a reasonable recycling experience exists within the US. It seems highly likely the recycling technology will continue to develop at a fast pace to support the worldwide mixed-oxide (MOX) fuel reprocessing system and expand the use of fission nuclear power. Fusion has a much longer timescale than 30 years and will certainly benefit from the ongoing fission recycling experience and related governmental regulations.

5.1.1. Reuse within Nuclear Industry

The concept of reuse of metal and concrete within the nuclear industry can be developed further. For example, the steel in reinforcing rods (called rebar) used in construction of concrete buildings is mildly radioactive. When encased in meter-thick concrete used for containment buildings, the emitted radiation from low radioactivity steel would pose a vanishingly small dose to any worker onsite. Hrncir [60] suggests that recycled steel could be used in the nuclear industry for the liner plates in rooms, or for components such as tanks, accumulators, etc. Concrete small rubble that poses < 1 mrem/year dose to an exposed worker could be used in construction of a new facility, such as for road or parking lot beds, or substrate under building foundations, or pads for equipment, etc.

The reuse of metals has been performed successfully in the US on a small scale within the nuclear industry and at national laboratories. Since the American scrap metal industry is highly concerned about radioactivity in their products [62], the process has been "constrained" or "restricted" to recycling only mildly radioactive materials back to the nuclear industry.

The DOE has operated small-scale constrained releases to the nuclear industry since the 1990s. Several cases of successful free release and constrained release recycling are described below:

- About 150 tons of copper from the magnets in the decommissioned 184-inch cyclotron at Lawrence Berkeley Laboratory had mild activation, reading between 0 and 20 pCi/g (0-0.7 Bq/g), with an average of 3 pCi/g (0.1 Bq/g), due to approximately 0.42 mCi (15 MBq) of Co-60 dispersed volumetrically in the copper. The California Department of Health Services considered this copper to be non-radioactive and permitted free release of this copper to the recycle market [63].
- At the INL, lead has been retrieved from shielding casks and recycled into both the nuclear industry and the commercial industry as free release material. In 1999, the INL identified, characterized, and shipped lead-and-steel casks to the GTS-Duratek firm in Oak Ridge,

Tennessee for recycling. GTS-Duratek recycled nearly 100 tons of lead from those casks on the open market. GTS-Duratek has also recycled over one hundred tons of mildly activated steel from the Belgian BR3 reactor and fabricated steel shielding bricks for use at DOE accelerator facilities [64]. Note that GTS-Duratek was purchased by the EnergySolutions company in 2006.

- The INL recycled more cask shielding with the GTS-Duratek firm in 2001 to fabricate another 100 tons of lead bricks for nuclear industry use as an accelerator target shielding wall at the Idaho Accelerator Center on the campus of Idaho State University. Each shielding brick weighed 26.25 pounds (11.9 kg). The estimated cost of LLW disposal of the lead was about $5 USD/pound while the approximate cost of recycling was $4.30 USD/pound, which included fabrication into brick shapes. Therefore, there was a cost saving in recycle versus disposal and the savings of disposal volume. The cost to purchase brand new shielding bricks rather than obtain recycled bricks was estimated to be $46.00 USD per brick, so there was the savings of not requiring the purchase of new bricks.
- Recycling stainless steel and lead has also been carried out within the nuclear industry at the INL. In 1996, a program to utilize mildly radioactive (surface and volumetric) stainless steel and mildly radioactive lead to fabricate small shielding containers sized to accommodate one 30-gallon or one 55-gallon waste drum was carried out and several prototype casks were fabricated [65]. The drums contained transuranic waste (TRU) that read on the order of 50 mSv/h (5 rems/h). The casks were designed, built, and tested for strength and impact [66,67]. After passing the tests, the casks were used to shield the drums to the INL 'hands-on' contact handling level of 2 mSv/h (200 mrem/h). The casks were put in service at the Radioactive Waste Management Complex at the INL. The prototype casks functioned well and are still in use.
- The Savannah River National Laboratory (SRNL) also experimented with recycle and beneficial reuse within the DOE community [68,69]. Some large parts, e.g., reactor coolant pumps from decommissioned production reactors, were melted and cast as ingots; these steel ingots were shaped appropriately for later use as shielding blocks. The SRNL also examined construction of TRU and HLW waste casks using recycled materials, and has fabricated some casks similar to those at INL [68].

- After the DOE placed a suspension on the release of scrap metal in 2000, the Duratek company investigated the reuse of contaminated lead (Pb) for use within the DOE. The 250 tons of lead came from several DOE sites and some Department of Defense facilities. The lead was melted and then recast for use in waste cask overpack shielding, cask shielding, and lead bricks [70].

Such experiences prove the feasibility of recycling metals within the nuclear industry. Noteworthy is that during the scrap metal melting process, the slag tends to collect some of, or a majority of, the radionuclides. When the slag is removed from the melt, the resulting ingots contain only very low levels of radioactivity. The slag would then be sent to LLW disposal, but as a greatly reduced volume. The tests with the INL shielding containers showed that millwright composition adjustments after slag removal in the foundry produced metal alloys with properties very similar to, or equal to, those of virgin alloys.

It has been noted that the process of melting metal alloys causes separation of some radionuclides, such as Co-60 that remains in solution while Cs-137 does not [37]. This separation should be investigated to determine if a facility dedicated to treating mildly activated waste was used to handle these materials, perhaps most of the radionuclides would be removed from the scrap metals, thus reducing their radioactivity. A facility such as the former Duratek plant in Tennessee could investigate melting separation of radioactive materials. If separation is feasible, then the small amounts of radionuclides separated from the scrap could be handled as LLW with the bulk of the metal being reclaimed for use. Investigation of means to scavenge radionuclides such as Co-60 should also be pursued.

Another issue is the recycling of concrete. The DOE has large amounts of used concrete for disposal, roughly estimated to be more than a million cubic meters. Several options were developed to assess their disposal within the DOE nuclear community. The main goal is to minimize the dose to members of the public and total cost. The concrete is profiled for radionuclides [71] and the corresponding dose of pCi/g per isotope present is summed and compared to a total 10 μSv/y (1 mrem/y) dose to members of the public. If the concrete dose is > 10 μSv/y, the DOE Assistant Secretary of Environment, Safety and Health would have to give approval for the project release limit, and if the dose is < 10 μSv/y the DOE Field Office with project responsibility can approve the project release limit. A consensus standard [72] has been written about concrete disposal options.

5.1.2. Technical Aspects of Recycling, Critical Issues, and Needs

Here, the technical feasibility of recycling is based on the dose rate to the RH equipment. Essentially, the dose determines the RH needs (hands-on, conventional, or advanced tools to handle the radioactive components) and the interim storage period necessary to meet the dose limit. Beside the recycling dose, other important criteria include the decay heat level during reprocessing, recycling of tritium-containing materials, physical properties of recycled products, and economics of fabricating complex shapes remotely.

Figure 7 displays the recycling dose for ARIES-ACT-2 components. The variation with time of the recycling dose shows a strong location dependence. The highly irradiated components can potentially be recycled in less than a year after shutdown with advanced RH equipment that can handle 10,000 Sv/h or more. The FW that surronds the plasma is an integral part of the blanket. It is shown in Figure 7 as a separate component to provide the highest possible dose to RH equipment. The average FW/blanket dose is an order of magnitude lower. ^{54}Mn (from Fe) is the main contributor to the dose of FS-based components (FW, blanket, SR and VV) at early cooling periods (<10 y), while impurities have no contribution to the recycling dose during such a period. Storing the FW/blanket temporarily for several years helps drop the dose by a few orders of magnitude before recycling. The outermost components (cryostat and bioshield) contain very low radioactivity and can be recycled with hands-on following a specific cooling period. We expect insignificant dose buildup due to the reuse of materials after numerous life cycles based on previous multiuse analysis performed for the divertor component [21].

Advanced RH equipment has been used in the nuclear industry, in hot cells and reprocessing plants, and in spent fuel facilities [73]. Reviews of remote procedures currently used within the nuclear industry suggest that the 0.01 Sv/h criteria for conventional RH equipment (10^3–10^4 times the hands-on dose limit) have been unduly conservative [10,11]. The re-melting of wastes from fission power plants has already been carried out on materials with a contact dose rate of 0.12 Sv/h [11]. Much higher dose rates are present in routine operations in the reprocessing of spent fuel in vitrification facilities. Contact dose rates of 3,000 – 10,000 Sv/h exist at the outside surfaces of cylinders during operations such as weighing, welding, cleaning, contamination monitoring, and transfer to flasks [10,74]. In the 1960s, the ANL-West Hot Fuel Examination Facility developed radiation resistant tools to handle fuel rods for the Experimental Breeder Reactor (EBR-II). The RH equipment operated successfully at 10,000 Sv/h. While the fission processes may have no direct relevance to fusion, their success gives confidence that advanced RH techniques could be developed to handle high doses (> 10,000 Sv/h) for the recycling of fusion materials.

Since the recycling approach plays an essential role in minimizing the volume of radwaste, it should be pursued despite the lack of detail on how to implement it at the present time. Based on known areas involved in the recycling process, we identified several key issues for the fusion community to examine with a dedicated R&D program:

Recycling issues:

- Separation of various activated materials from complex components
- Radiochemical or isotopic separation processes for some materials, if needed
- Treatment and remote re-fabrication of radioactive materials
- Radiotoxicity and radioisotope buildup and release by subsequent reuse
- Properties of recycled materials? Any structural role? Reuse as filler?
- Handling of T containing materials during recycling
- Management of secondary waste. Any materials for disposal? Volume? Radwaste level?
- Energy demand for recycling process
- Cost of recycled materials
- Recycling plant capacity and support ratio.

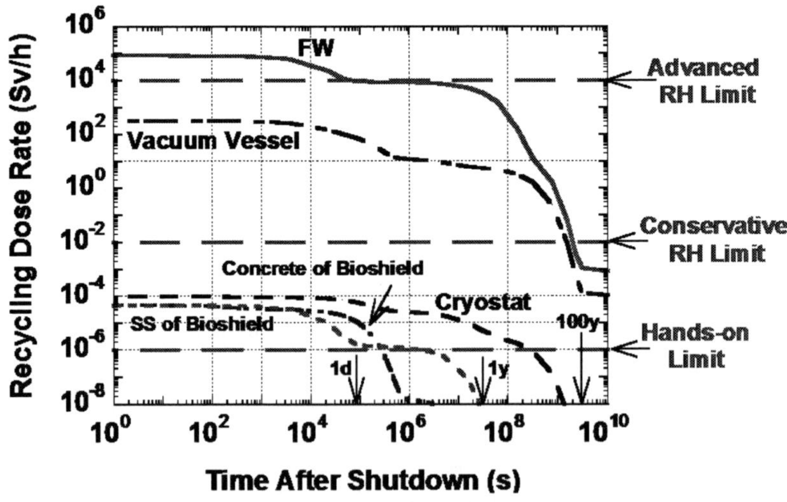

Figure 7. Recycling dose for selected ARIES-ACT-2 components.

Recycling needs:

- Radiation-resistant remote handling equipment
- Reversible assembling process of components and constituents (to ease separation of materials after use)
- Efficient detritiation system to remove tritium before recycling
- Large and low-cost interim storage facility with decay heat removal capacity
- Nuclear industry should accept recycled materials
- Recycling infrastructure.

5.2. Clearance

The majority of nuclear waste (> 70%) contains traces of radioactive nuclides that represent no risk to the public safety and health. Over the past several decades, researchers in the nuclear field have attempted to issue policy statements to deregulate such materials with low concentrations of radioactive contamination. If this effort succeeds, the clearable materials will not be subject to regulatory control, be handled as if they are no longer radioactive, and be unrestrictedly recycled into consumer products (tables, chairs, tools, building and road materials, etc.), saving high disposal costs and freeing ample space in repositories for more radioactive waste. Issuing an official clearance policy by NRC in the near future is extremely valuable, as the decommissioning of nuclear plants will generate a huge amount of slightly radioactive materials. Concrete constitutes the largest volume of slightly radioactive solids resulting from decommissioning.

5.2.1. Release of Slightly Radioactive Materials to Commercial Market

In the US, clearance, or release from regulatory control to the open market (i.e., free release) for volumetrically activated or contaminated materials, has been performed only on a case-by-case basis during decommissioning projects since the 1990s. There is no established market for clearable materials and, as will be discussed later, there is considerable reluctance from the US commercial sector to begin recycling these free release materials. Despite this situation, there have been some steps forward in clearance:

- The Health Physics Society and the ASTM have published guidance on clearance [75,76] that focuses on a dose limit of less than 10 μSv/y

(1 mrem/y) to any individual member of the public from these materials

- NRC issued the NUREG-1640 document [12] that gives additional direction on calculating potential annual doses from released material for a wide number of exposure scenarios.

Thus far, the US has had some very small successes with free release, as mentioned below, and more success with 'restricted release' of clearable materials to be re-used within the nuclear industry or the DOE community. In this category, the slightly radioactive materials are not recycled into a consumer product, but rather released to dedicated nuclear-related facilities under continuing regulatory control or to specific applications where contact for exposure of the general public is minimal. Examples include shielding blocks for containment buildings of licensed nuclear facilities, concrete rubble base for roads, deep concrete foundations, non-water supply dams for flood control, etc.

5.2.2. Historical Evolution of Clearance Standards

Since the 1940s, the US Atomic Energy Commission (AEC) and its successor agency, the NRC, made several attempts to set standards for release of slightly radioactive materials from regulatory control for licensed US facilities such as fission power reactors, fuel fabrication and reprocessing plants, accelerators, hospitals, etc. During the decade of the 1970s, the NRC gave greater uniformity to the clearance standards while materials containing traces of radioactivity continued to be released to date using the case-by-case approach. More attempts by the NRC in 1980, 1990, 1998, and more recently in 2003 declared materials with low concentrations of radioactivity can be deregulated. The 1998 draft NUREG-1640 document [77] contains estimates of the total effective dose equivalent (from which the clearance index can be derived) for 67 radionuclides that could be present in metals from decommissioning of nuclear facilities. The draft NUREG-1640 document has been published in a final form [12] in 2003 for 115 radioisotopes that can be found in four types of clearable materials: steel, copper, and aluminum scrap, and concrete rubble. This document makes no distinction of surface or volumetric contamination, just doses to exposed persons from cleared materials.

The 2003 NUREG-1640 document has been used to support US regulatory considerations. However, this technical report cannot be inferred to represent any US regulatory decision as the NRC has not yet issued an official policy on the unconditional release of specific materials. Herein, the proposed annual doses reported in the NUREG-1640 documents [12,77] are used to obtain the clearance indices (CI) for the 115 radioisotopes and referred to as the proposed US CI limits. Even though we are content to only use the proposed US CI limits in our power plant analysis, it is pertinent to mention the DOE 2002 guide [78] that has been used to control the recycling and reuse of slightly radioactive materials from numerous DOE sites not licensed by the NRC. This guide was archived in 2011, but portions of this guide may be incorporated into future DOE technical standards on waste and recycling. Some DOE facilities are no longer functioning but still contain significant amounts of slightly radioactive materials.

In a series of documents issued in the early 1980s and continued through the 2000s, the International Atomic Energy Agency (IAEA) established the principles that underlie its technical estimates of the clearance limits. In 1996, the IAEA prepared an interim report on recommended clearance limits for solid materials [79] for 1650 radionuclides of interest to fission and fusion applications. In 2004, the IAEA published revised clearance standards [13] for 257 radionuclides, claiming to take into account the US NUREG-1640 [12] evaluation.

There is a widespread agreement between the US NRC and IAEA organizations on the primary dose standard and the negligible risk the cleared materials present to individuals. Both organizations recommended an individual dose standard of 10 μSv/y (1 mrem/y) for cleared solids that is < 1% of the radiation received each year from natural background sources. Nevertheless, we observed a notable difference between the clearance limits developed by the US NRC [12] and IAEA [13]. Furthermore, numerous fusion radioisotopes with $T_{1/2} \geq 10$ y are missing and should be included in future evaluations. These include, but are not limited to, ^{10}Be, ^{26}Al, ^{32}Si, ^{91,92}Nb, ^{98}Tc, ^{113m}Cd, ^{121m}Sn, ^{150}Eu, ^{157,158}Tb, ^{163,166m}Ho, ^{178n}Hf, ^{186m,187}Re, ^{193}Pt, ^{208,210m,212}Bi, and ^{209}Po. Figure 8 displays the ratio of the US NRC limits to the IAEA's. As noticed, both standards do not agree on the limits for many radioisotopes because different approximations are used to compute these limits and different exposure scenarios are selected to model the doses. Consistency on the clearance standards is highly desirable to preclude issues of US receiving scrap metal from abroad that has higher-than-allowable radioactivity. The US scrap metal industry and the public continue to have concerns with this particular issue, as recent Nuclear Materials Events Database [80] showed that imported scrap metal and metal consumer products from other countries (most notably Mexico, China, and India) are sometimes mildly radioactive. It is fortunate the US Customs and metal recycling companies perform surveys for radioactivity.

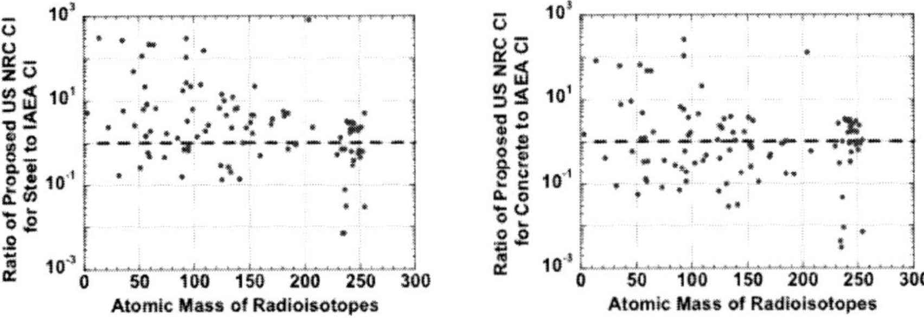

Figure 8. Ratio of 2003 US NRC CI to 2004 IAEA CI for steel scrap and concrete rubble.

5.2.3. Societal Aspects of Clearance

There is considerable reluctance from the US commercial sector to begin recycling the clearable materials. The reluctance stems from the concern that US consumers will not purchase products that could possibly be 'tainted' by radioactivity, no matter how small the radioactivity. Even though the doses would be minor, less than microrems, a big part of public fear is that they are unaware of the risk they are accepting. Chen [81] pointed out that releasing mildly radioactive materials from regulatory control to the commercial metals industry could result in these materials finding their way into consumer products (e.g., strollers, toys, etc.). Other opposition groups argue that no one can guarantee a public dose of only 10 μSv/y (1 mrem/y) from a batch of recycled metals since it is not possible to know which products nor how many consumer products a person could purchase and keep in their home. These views have engendered negative perceptions among the materials industries due to potentially negative economic impacts of US consumer fear of the items and refusal to purchase the items. For these reasons, the US recycling market for clearable materials is presently very small.

The European experience is of interest. Beginning in the early 1990s, several European countries have issued regulations for clearance and several projects have cleared materials in industrial quantities, mostly metals and concrete rubble [16]. For instance:

- **Sweden** was among the first to apply clearance on metals and installed a "nuclear" furnace in its research center Studsvik. The Swedish regulation allows clearing of materials, but also allows treating and clearing metals from abroad, providing that material can be free released within 10 years. The decay storage and subsequent release are being carried out at the site of Studsvik.
- **Spain** started industrial clearance of metallic material with its Vandellos decommissioning project. An agreement was signed between the Spanish government and the steel industry to enhance the acceptance of clearable materials by steel recyclers.
- **Belgium** has introduced clearance levels into its regulations and has already cleared thousands of tons of steel and concrete from the BR3 and Eurochemic projects.

5.2.4. Example of Successful Free Releases

In the course of nuclear fission reactor equipment upgrades as well as decommissioning, sometimes materials are successfully released to the scrap metal market using the case-by-case approach. An example is the 2011 main condenser module replacement at the Columbia Generating Station in Washington. The condenser modules had radiological contamination in the tubes. The Perma-Fix Environmental Services Company decontaminated the condenser modules to keep them out of landfills and to realize some cost savings by recycling the metal [82]. The weights of these steel components are given below:

Condenser modules	1,197,484 kg
Water boxes	94,935 kg
Scrap metal from dismantlement	270,489 kg.

This equipment was decontaminated with a liquid process, generating 24,703 kg of liquid waste for disposal. Each condenser tube was surveyed to verify it was below these surface contamination release levels: 5,000 dpm/100 cm^2 total beta-gamma contamination, 100 dpm/100 cm^2 total alpha contamination, and 1,000 dpm/100 cm^2 for the Sr-90 radionuclide. These surface contamination levels are the average contamination values given in US NRC Regulatory Guide 1.86 [33]. The condenser tubes had to be sectioned in 3.7 m lengths to facilitate the free release survey. Additionally, 10% of the tubes were selected at random to be split open and surveyed by hand as a quality control check of the automated radiation counter. The 1,290 tonnes of steel was recycled in Washington by the Perma-Fix Company.

Another example is the Decontamination for Decommissioning (DFD) process that was used at the Big Rock Point and Maine Yankee power plants [83]. This process uses fluoroboric acid to remove outer scale and a thin layer of base metal so that fixed contamination is removed and the metal is reduced to below levels that allow free release. This process was successful in treating the primary coolant systems, including the pumps and heat exchangers, at these two power plants.

5.2.5. Technical Aspects of Clearance, Critical Issues, and Needs

The CI for a clearable material is the ratio of the activity (in Bq/g) of the individual radioisotope to the allowable clearance limit summed over all radioisotopes. A material qualifies for clearance if the CI drops below one at any time during a defined storage period following decommissioning. For components with multiple materials, it is essential to segregate and re-evaluate the CIs for constituents. Sizable components, such as the 2 m thick bioshield, should be segmented into small layers with the CI for each layer reexamined.

For the ARIES-ACT-2 design, the CIs for all internal components (blanket, SR, VV, shield) exceed unity by a wide margin even after an extended period of 100 y (refer to Figure 9). Nb-94 (from Nb impurity) is the main contributor to the CI of FS-based materials at 100 y. Interestingly, a VV made of pure iron is still not clearable (due to Co-60 generated by Fe). This means no matter how serious an attempt is to strictly control all impurities or alter the alloying elements of FS, the VV will never qualify for clearance, but remains recyclable. Even though the magnet is well protected by the blanket, VV, and shield, not all constituents are clearable. Only the Cu stabilizer and outer coil case can be cleared within 100 y.

The 2 m thick external concrete building (bioshield) that surrounds the tokamak represents the largest single component of the decommissioned materials. Fortunately, the bioshield along with the 40 cm thick cryostat (20% steel and 80% void) and some magnet constituents (Cu and outer coil case) qualify for clearance, representing ~70% of the total volume of radioactive materials. We divided the thick bioshield into four segments (0.5 m each) and reevaluated the CIs for its constituents (85% Type-04 ordinary concrete, 10% mild steel, and 5% He by volume). The innermost segment is clearable after shutdown (see Figure 9) while the outer segments exhibit CI $<<$ 1.

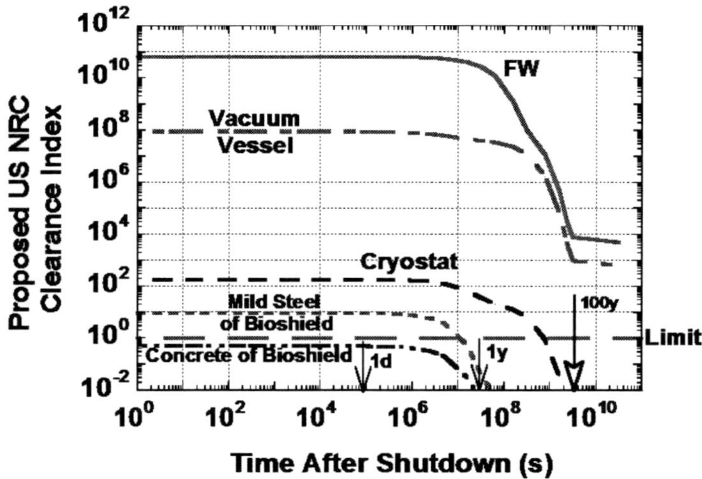

Figure 9. Reduction of clearance index with time after shutdown for selected ARIES-ACT-2 components.

The missing fusion-related radioisotopes from US NRC and IAEA CI evaluations make it difficult to claim with certainty the clearability of fusion materials. Efforts by the US NRC, IAEA, and other organizations should continue to develop clearance standards for all radioisotopes of interest to fusion applications. We support the national and international organizations' efforts to convince industrial as well as environmental groups that clearance of slightly radioactive solids can be conducted safely with no risk to the public health. Other clearance issues and needs that require further assessment include:

Clearance issues:

- Discrepancies between proposed US-NRC and IAEA clearance standards
- Impact on clearance index prediction of missing radioisotopes
- Radioisotope buildup and release by subsequent reuse.

Clearance needs:

- Official clearance limits issued by legal authorities
- Accurate measurements and reduction of impurities that deter clearance of in-vessel components
- Reversible assembly process of components and constituents

- Large and low-cost interim storage facility
- Clearance infrastructure
- Clearance market.

6. Final Remarks

The integration of the recycling and clearance processes in fusion designs is at an early stage of development. Due to the lack of experience, it is almost impossible to state how long it will take to refabricate fusion components out of mildly radioactive materials. The minimum time that one could expect is one-year temporary storage and two years for fabrication, assembly, inspection, and testing. All processes must be performed remotely with no personnel access to fabrication facilities. Figure 10 depicts the essential elements of the recycling/clearance process. Predicting the various steps, one could envision the following:

1. After extraction from the fusion power core, components are taken to the Hot Cell to disassemble and remove any parts that will be reused, separate into like materials, detritiate, and consolidate into a condensed form.
2. Ship materials to a temporary onsite or centralized facility to store for a period of ~1 year or less.

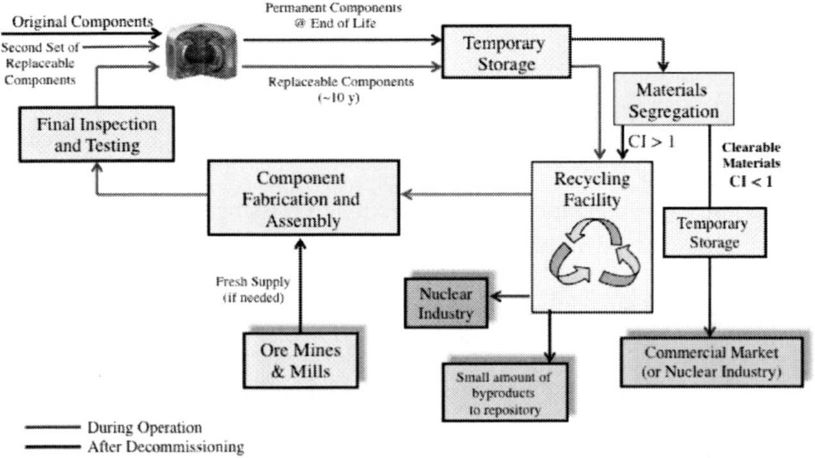

Figure 10. Diagram of recycling and clearance processes.

3. If the CI will not reach unity in less than 100 y, transfer materials to a recycling center to refabricate into useful forms. Fresh supply of materials could be added as needed.
4. If the CI will drop below unity in less than 100 y, store materials for a specific period, then release to the public sector to reuse without restriction.

It is just a matter of time to develop the recycling/clearance technology and official regulations. At present, the experience with recycling/clearance is limited, but will be augmented significantly by advances in fission reactor dismantling, used fuel reprocessing, and bioshield clearing before fusion is committed to commercialization in the 21st century. While there is no US official regulation for recycling and clearance of radioactive materials, there has been some progress made.

For instance:

- Limited scale recycling within the nuclear industry has been proven feasible at several US national laboratories
- Remote handling equipment operated well at high doses around 10,000 Sv/h since the 1960s
- The free release of clearable materials has been performed since the 1990s on a case-by-case basis during decommissioning projects
- Recently in 2010, the DOE required decontamination of 15,300 tons of radioactive nickel and recycling into products that will be used in radiologically-controlled applications
- MOX fuel fabrication facility in South Carolina is > 50% complete and planned to start operation in 2019.

Internationally, numerous fission industries are currently developing advanced techniques for spent fuel reprocessing while several regulatory agencies have issued guidelines for clearance. Such developments at the national and international levels will be of great importance to fusion, but adaptation is necessary to the fusion needs (radiation level, component size, weight, etc.).

Conclusion

Fusion should be able to avoid creating HLW since there is design latitude in fusion materials selection. However, the fusion LLW amounts are large, so efforts to recycle and clear are essential to support fusion deployment, reclaiming of resources (through less mining ore), and maintaining lifestyle in a technologically advanced society with minimal environmental impact.

Past fusion studies indicated recycling and clearance are technically feasible for any fusion device employing low-activation materials, using advanced radiation-resistant remote handling equipment, and having clearance guidelines for slightly radioactive materials. Nevertheless, such approaches are relatively easy to envision and apply from a science perspective, but could be a challenge from policy, regulatory, and public acceptance perspectives. In the US, the public continues to have fear of radiation exposure, which limits the possibility of clearance of slightly radioactive materials on a large scale. The desire to reclaim valuable resources in the form of metal alloys and concrete rubble persists in the nuclear industry. The efforts by government agencies to create paths forward have come under scrutiny by the public, watchdog groups, and the metal recycling industry. The issue of clearance remains a point of contention between the nuclear industry and these other groups. To reverse this situation and make the recycling and clearance approaches a reality, major rethinking, education, and research should be developed and pursued. In the meantime, the fusion program should be set up to accommodate the recycling/clearance strategy to continue holding the promise of fusion energy production with low environmental impact.

Acknowledgments

The authors would like to acknowledge the support of this work by their institutions: the Fusion Technology Institute at the University of Wisconsin-Madison and the Fusion Safety Program at the Idaho National Laboratory.

References

[1] Ponti, C. (1988). “Recycling and shallow land burial as goals for fusion reactor materials development,” *Fusion Technology*, *13*, 157-164.

[2] Cheng, E. T., Sze, D. K., Sommers, J. A. & Farmer III, O. T. (1992). "Materials recycling considerations for D-T fusion reactors," *Fusion Technology*, *21*, 2001-2008.

[3] Dolan, T. J. & Butterworth, G. J. (1994). "Vanadium recycling," *Fusion Technology*, *26*, 1014-1020.

[4] Rocco, P. & Zucchetti, M. (1998). "Advanced management concepts for fusion waste," *Journal of Nuclear Materials*, *258-263*, 1773-1777.

[5] Petti, D., McCarthy, K., Taylor, N., Forty, C. B. A. & Forrest, R. A. (2000). "Re-evaluation of the use of low activation materials in waste management strategies for fusion," *Fusion Engineering and Design*, *51-52*, 435-444.

[6] El-Guebaly, L., Henderson, D., Abdou, A. & Wilson, P. (2001). "Clearance issues for advanced fusion power plants," *Fusion Technology*, *39*, 986-990.

[7] Zucchetti, M., Forrest, R., Forty, C., Golden, W., Rocco, P. & Rosanvallon, S. (2001). "Clearance, recycling and disposal of fusion activated material," *Fusion Engineering and Design*, *54*, 635-643.

[8] Tobita, K., Nishio, S., Konishi, S. & Jitsukawa, S. (2004). "Waste management for JAERI fusion reactors," *Journal of Nuclear Materials*, *329-333*, 1610-1614.

[9] El-Guebaly, L., Wilson, P., Henderson, D. & Varuttamaseni, A. (2004). "Feasibility of target materials recycling as waste management alternative," *Fusion Science and Technology*, *46*, 506-518.

[10] El-Guebaly, L., Forrest, R. A., Marshall, T. D., Taylor, N. P., Tobita, K. & Zucchetti, M. (2005). "Current challenges facing recycling and clearance of fusion radioactive materials," University of Wisconsin Fusion Technology Institute Report, UWFDM-1285, Available at: http://fti.neep.wisc.edu/pdf/fdm1285.pdf.

[11] Zucchetti, M., El-Guebaly, L., Forrest, R., Marshall, T., Taylor, N. & Tobita, K. (2007). "The feasibility of recycling and clearance of active materials from a fusion power plant," *Journal of Nuclear Materials*, *367-370*, 1355-1360.

[12] Anigstein, R. et al., "Radiological Assessments for Clearance of Materials from Nuclear Facilities," volume 1, NUREG-1640, US Nuclear Regulatory Commission, June 2003. Available at: http://www.nrc.gov/reading

[13] International Atomic Energy Agency, Application of the concepts of exclusion, exemption and clearance, IAEA Safety Standards Series, No.

RS-G-1.7 (2004). Available at: http://www.pub.iaea.org/MTCD/publications/PDF/Pub1202_web.

[14] El-Guebaly, L., Wilson, P. & Paige, D. (2006). "Evolution of clearance standards and implications for radwaste management of fusion power plants," *Fusion Science and Technology*, *49*, 62-73.

[15] El-Guebaly, L. (2007). "Evaluation of disposal, recycling, and clearance scenarios for managing ARIES radwaste after plant decommissioning," *Nuclear Fusion*, *47*, 485-488.

[16] El-Guebaly, L., Massaut, V., Tobita, K. & Cadwallader, L. (2008). "Goals, Challenges, and Successes of Managing Fusion Active Materials," *Fusion Engineering and Design*, *83*, 928-935.

[17] Massaut, V., Bestwick, R., Brodén, K., Di Pace, L., Ooms, L. & Pampin, R. (2007). "State of the art of fusion material recycling and remaining issues," *Fusion Engineering and Design*, *82*, 2844-2849.

[18] Zucchetti, M., Di Pace, L., El-Guebaly, L., Kolbasov, B. N., Massaut, V., Pampin, R., & Wilson, P. (2009). "The Back End of the Fusion Materials Cycle," *Fusion Science and Technology*, *52*, 109-139.

[19] El-Guebaly, L. A. & Cadwallader, L. (2009). "Recent Developments in Safety and Environmental Aspects of Fusion Power Plants." Chapter 9 in book: Nuclear Reactors, Nuclear Fusion and Fusion Engineering. A. Aasen and P. Olsson Editors. NOVA Science Publishers, Inc.: Hauppauge, New York, USA. ISBN: 978-1-60692-508-9, 321-365.

[20] Di Pace, L., El-Guebaly, L., Kolbasov, B., Massaut, V. & Zucchetti, M., (2012). "Radioactive Waste Management of Fusion Power Plants," Chapter 14 in Book: Radioactive Waste. Dr. Rehab Abdel Rahman (Ed.), ISBN: 978-953-51-0551-0, InTech Available at http://www.intechopen.com/books/radioactive-waste.

[21] Kolbasov, B., El-Guebaly, L., Khripunov, V., Someya, Y., Tobita, K. & Zucchetti, M. "Some Technological Problems of Fusion Materials Management," To be published in Fusion Engineering and Design. Available online at: http://authors.elsevier.com/sd/article/S09203796-14000453

[22] EL-Guebaly, L., et al., "Neutronics and Shielding Characteristics of ARIES-ACT-2 Power Plant with DCLL Blanket," University of Wisconsin Fusion Technology Institute Report, UWFDM-1418 (2014). Available at: http://fti.neep.wisc.edu/pdf/fdm 1418.pdf.

[23] Zinkle, S. J., Möslang, A., Muroga, T. & Tanigawa, H. (2013) "Multimodal options for materials research to advance the basis for fusion energy in the ITER era," *Nuclear Fusion*, *53*, No. 10, 104024.

[24] Zinkle, S. J. (2013). "Challenges in developing materials for fusion technology - past, present and future," *Fusion Science and Technology*, *64*, 65-75.

[25] El-Guebaly, L., Kurtz, R., Rieth, M., Kurishita, H. & Robinson, A. (2011). "W-Based Alloys for Advanced Divertor Designs: Options and Environmental Impact of State-of-the-Art Alloys." Fusion Science and Technology, *60*, 185-189.

[26] The ITER Project: http://www.iter.org.

[27] El-Guebaly, L. & Mynsberge, L. "Neutronics Characteristics and Shielding System for ARIES-ACT-1 Power Plant," Fusion Science and Technology, in press.

[28] Maisonnier, D., Cook, I., Sardain, P., Aandreani, R., di Pace, L., Forrest, R. et al., (2005). "A conceptual study of commercial fusion power plants." Final Report of the European Fusion Power Plant Conceptual Study (PPCS), Report EFDA-RP-RE-5.0.

[29] The ESBWR Design: http://www.gepower.com/prod_serv/products/nuclear_energy

[30] US Code of Federal Regulations, Title 10, Energy, Part 60, Disposal of High Level Radioactive Wastes in Geologic Repositories, US Government Printing Office, January 2014.

[31] US Code of Federal Regulations, Title 10, Energy, Part 61, Licensing Requirements for Land Disposal of Radioactive Waste, US Government Printing Office, January 2014.

[32] US Code of Federal Regulations, Title 10, Energy, Part 71, Packaging and Transportation of Radioactive Material, US Government Printing Office, January 2014.

[33] Termination of Operating Licenses for Nuclear Reactors, Regulatory Guide 1.86, *US Atomic Energy Commission*, June 1974.

[34] Banovac, K., Buckley, J., Hickman, J., Persinko, A., Shepard, J., Smith, T. & Watson, B. (2010). "Power Reactor Decommissioning – Regulatory Experiences from Trojan to Rancho Seco and Plants In-Between," Proceedings of the ANS Topical Meeting on Decommissioning, Decontamination, and Reutilization (DD&R 2010), Idaho Falls, Idaho, August 29-September 2, 2010, American Nuclear Society.

[35] US Code of Federal Regulations, Title 10, Energy, Part 834, Radiation Protection of the Public and the Environment, Section 301, Release of property containing residual radioactive material, US Government Printing Office, January 2014.

[36] Turner, R. (2006). "Scrap Metals Industry Perspective on Radioactive Materials," *Health Physics*, *91*, 489-493.

[37] Lubenau, J. O. & Yusko, J. G. (1995). "Radioactive Materials in Recycled Metals," *Health Physics*, *68*, 440-451.

[38] Lubenau, J. O. & Yusko, J. G. (1998). "Radioactive Materials in Recycled Metals – an Update," *Health Physics*, *74*, 293-299.

[39] Draft Programmatic Environmental Assessment for the Recycle of Scrap Metals Originating from Radiological Areas, DOE/EA-1919, US Department of Energy, December 2012.

[40] Radiation Protection of the Public and the Environment, DOE Order 458.1, US Department of Energy, February, 11, 2011.

[41] Hammer, D., Sullivan, N., Zatz, M., Woods, H. & Arcaya, M. (2004). Summary and Categorization of Public Comments on Controlling the Disposition of Solid Materials, NUREG/CR-6682, supplement 1, US Nuclear Regulatory Commission, March.

[42] US Code of Federal Regulations, Title 10, Energy, Part 61, Licensing Requirements for Land Disposal of Radioactive Waste, Section 55, Waste classification, US Government Printing Office, January 2014.

[43] Tang, Y. S. & Saling, J. H. (1990). Radioactive Waste Management, Hemisphere Publishing Corporation, New York, chapter 6.

[44] Annual Transuranic Waste Inventory Report – 2013, DOE/TRU-13-3425, revision 1, US Department of Energy, February 2014.

[45] Radiological Release Event at the Waste Isolation Pilot Plant on February 14, 2014, Accident Investigation Report Phase 1, US Department of Energy, April 2014. Available at: http://energy

[46] The use of scientific and technical results from underground research laboratory investigations for the geological disposal of radioactive waste, IAEA TECDOC-2143, International Atomic Energy Agency, Vienna, Austria, September 2001.

[47] Preuss, J., Eilers, G., Mauke, R., Muller-Hoeppe, N., Engelhardt, H. J., Kreienmeyer, M., Lerch, C., Schrimpf, C. (2002). "Post Closure Safety of the Morsleben Project," *Proceedings of the Waste Management 2002 Conference (WM'02)*, February 24-28, Tucson, Arizona.

[48] Garrick, B. J. et al., (2009). Survey of National Programs for Managing High-Level Radioactive Waste and Spent Nuclear Fuel, a report to Congress and the Secretary of Energy, US Nuclear Waste Technical Review Board, Arlington, VA, October, available at: http://www.nwtrb.gov/reports/reports.html

[49] Geological Disposal, How the World is Dealing with its Radioactive Wastes, United Kingdom Nuclear Decommissioning Authority, ISBN 978-1-905985-32-6, 2013.

[50] Garrick, B. J. et al., (2011). Experience Gained from Programs to Manage High-Level Radioactive Waste and Spent Nuclear Fuel in the United States and Other Countries, a report to Congress and the Secretary of Energy, US Nuclear Waste Technical Review Board, Arlington, VA, April, available at: http://www.nwtrb.gov/reports/reports.html

[51] Taboas, A. L., Moghissi, A. A., LaGuardia, T. S. (2004). The Decommissioning Handbook, American Society of Mechanical Engineers, New York, chapter 19.

[52] Affordable Cleanup?, National Academy Press, Washington, DC, 1996, p 60.

[53] Smith, K. "Mission complete," Construction & Demolition Recycling, volume 15, number 1, January/February 2013, pages 14-18. Available at: http://www.cdrecycler.com/digital//20130102/index.html

[54] Devgun, J. S. (2002). "Impact of Lack of Consistent Free Release Standards on Decommissioning Projects and Costs," *Proceedings of the Waste Management 2002 Conference (WM'02)*, February, 24-28, Tucson, Arizona.

[55] Underground Salt Haul Truck Fire at the Waste Isolation Pilot Plant, February 5, 2014, Accident Investigation Report, US Department of Energy, March 2014. Available at: http://energy

[56] Report on Waste Burial Charges, Changes in Decommissioning Waste Disposal Costs at Low-Level Waste Burial Facilities, NUREG-1307, revision 15, US Nuclear Regulatory Commission, January, 2013.

[57] Smith, R. I., et al., (1988). Technology, Safety and Costs of Decommissioning a Reference Pressurized Water Reactor Power Station, NUREG/CR-0130, addendum 4, *US Nuclear Regulatory Commission*, July.

[58] Oak, H. D., et al., (1990). Technology, Safety and Costs of Decommissioning a Reference Boiling Water Reactor Power Station, NUREG/CR-0672, Addendum 4, US Nuclear Regulatory Commission, December.

[59] Fetter, S., Cheng, E. T., Mann, F. M. (1990). "Long term radioactive waste from fusion reactors: part II," *Fusion Engineering and Design*, *13*, 239-246.

[60] Hrncir, T., et al., (2013). "The impact of radioactive steel recycling on the public and professionals," *Journal of Hazardous Materials*, *254-255*, 98-106.

[61] Bartenev, S. A., Kvasnitskij, I. B., Kolbasov, B. N., Romanov, P. V., Romanovskij, V. N. (2004). "Radiochemical reprocessing of V-Cr-Ti alloy and its feasibility study," *Journal of Nuclear Materials*, *329-333*, 406-410.

[62] Sandoval, D. (2003). "Casting a nuclear shadow," Recycling Today, *41*, No. 12 (December), 42-46.

[63] U.S. Department of Energy, Environmental assessment for the slightly activated copper coil windings from the 184-inch cyclotron at Lawrence Berkeley Laboratory, Berkeley, California, DOE/EA-0851 (1993).

[64] Klein, M., Dadoumont, J., Demeulemeester, Y. & Massaut, V. (2001). "Experience in decommissioning activities at the BR3 site," *Fusion Engineering and Design*, *54*, 443-449.

[65] Richins, W. D., Fewell, T. E., Welland, H. J. & Sheely, H. R. (2000). Jr., Shielded containers for radioactive waste using recycled contaminated metals, *Nuclear Engineering and Design*, *197*, 183-195.

[66] Richins, W. D. (1996). Impact and structural analysis of the INEL 30 gallon recycled shielded storage container, *Idaho National Laboratory report*, INEL-96/0141.

[67] Richins, W. D. (1996). Impact and structural analysis of the INEL 55 gallon recycled shielded storage container, *Idaho National Laboratory report*, INEL-96/0272.

[68] Boettinger, W. L. (1993). "Beneficially reusing low level regulated waste, the Savannah River Site Stainless Steel Program," CONF-931207-1, Westinghouse Savannah River Company.

[69] Boettinger, W. L. & Mishra, G. (1997). "Stainless steel RSM beneficial reuse technical feasibility to business reality," WSRC-RP-97-292, Westinghouse Savannah River Company.

[70] Reno, C. (2003). "Progress in Recycling Elemental Lead for Reuse of Radiologically-Contaminated with the Nuclear Industry," *Proceedings of the Waste Management 2003 Conference (WM'03)*, February 23-27, Tucson, Arizona.

[71] Aggarwal, S., Charters, G. & Thacker, D. (2003). "Building material characterization using a concrete floor and wall contamination profiling technology," *Proceedings of Waste Management 2003 Symposium (WM'03)*, February 23-27, Tucson, Arizona.

[72] Standard guide for evaluating disposal options for concrete from nuclear facility decommissioning, ASTM E2216-02, American Society for Testing and Materials, West Conshohocken, PA, (2002).

[73] Desbats, P., Toubon, H. & Piolain, G. (2004). "Status of development of remote technologies applied to Cogema spent fuel management facilities in France," ANS 10th International Topical Meeting on Robotics and Remote Systems for Hazardous Environments, Gainesville, Florida.

[74] Pampin, R., Forrest, R. A. & Bestwick, R. (2006). Consideration of strategies, industry experience, processes and time scales for the recycling of fusion irradiated material, UKAEA report FUS-539.

[75] Surface and volume radioactivity standards for clearance, HPS/ANSI N13.12, Health Physics Society, McLean, Virginia (1999).

[76] Standard guide for unrestricted disposition of bulk materials containing residual amounts of radioactivity, ASTM E1760-96 (reapproved 2003), ASTM International, West Conshohocken, PA (1996).

[77] Radiological Assessments for Clearance of Equipment and Materials from Nuclear Facilities, Draft NUREG-1640, Nuclear Regulatory Commission, Washington, D.C. (1998).

[78] Control and Release of Property with Residual Radioactive Material, Guide, DOE G 441.1-XX, US Department of Energy, Washington, D.C. (2002).

[79] Clearance Levels for Radionuclides in Solid Materials – Application of Exemption Principles, Interim Report IAEA-TECDOC-855, International Atomic Energy Agency, Vienna (1996).

[80] US Nuclear Regulatory Commission, Nuclear Materials Events Database, https://nmed.inl.gov/, password required, accessed May 2014.

[81] Chen, S. Y. (2006). "Managing the disposition of potentially radioactive scrap metal," *Health Physics*, *91*, 461-469.

[82] Polley, G. M. (2012). "Packaging, Transportation and Recycling of NPP Condenser Modules," *Proceedings of the Waste Management 2012 Conference (WM'12)*, February 26 – March 1, Phoenix, Arizona.

[83] Bushart, S. P., Wood, C. J., Bradbury, D. & Elder, G. (2007). "Decontamination and Recycling of Radioactive Materials from Retired Components," *Proceedings of the Waste Management 2007 Conference (WM'07)*, February 25 – March 1, Tucson, Arizona.

In: Radioactive Waste
Editor: Susanna Fenton
ISBN: 978-1-63321-731-7

Chapter 2

THE CANISTER QUANDARY: SPENT NUCLEAR FUEL PACKAGING CONCEPTS FOR STORAGE, TRANSPORTATION AND DISPOSAL IN THE U.S.

Ernest L. Hardin
Sandia National Laboratories, Albuquerque, NM, US

ABSTRACT

Commercial spent nuclear fuel (SNF) is accumulating in the U.S. at the rate of approximately 2,000 metric tons (MT) per year, and a similar amount is being transferred each year from power plant fuel pools to dry storage. Most of the welded, sealed canisters used for dry storage can also be used for eventual transport to centralized storage or a geologic repository, hence these are called dual-purpose canisters (DPCs). By the year 2035 approximately half of all SNF in the U.S. is projected to be stored in DPCs. Unless these can be disposed of directly, they will all eventually need to be cut open and the SNF waste packaged in new containers.

DPC design has changed since they came into commercial use 20 years ago: they are larger and use various means to control criticality. Changes have occurred in parallel with advances in thermal and criticality analyses. Disposal has not been a factor in DPC design, due partly to the terms of contracts between utility companies and the government, which must ultimately disposition the SNF. A new canister design that is suitable for disposal also, has been addressed on two previous occasions: the multi-purpose canister (MPC) initiative of the 1990's, and the 2008 transport-aging-disposal (TAD) canister concept developed for a proposed repository in volcanic tuff. Both concepts were abandoned, hence SNF continues to accumulate in DPCs which are not purpose-designed or licensed for disposal.

The disposability of SNF canisters depends on effective safety strategies that ensure waste will be isolated in the repository, with sufficient heat dissipation and control of criticality. Such strategies will depend on disposal site characteristics, which may vary significantly for different host geologic media and disposal concepts. Over many thousands of years waste packages will cool, but will eventually be damaged by corrosion and may fill with ground water leading to the possibility of criticality. A framework for canister disposability that takes into account alternative safety strategies shows how repository programs in the U.S. and other countries have proceeded. It also shows what canisters options are available in the U.S. as alternatives to DPCs. Direct disposal of DPCs is also an option that is under study.

The potential advantages of canisters that are suitable for storage, transportation and disposal, or of direct disposal of existing DPCs, include simpler SNF management, lower cost, less secondary waste (e.g., used DPC hulls), and less worker exposure. Estimated cost savings in the U.S. depend on when such measures are introduced. Timely implementation is projected to save on the order of $10 billion in re-packaging costs. The capability to dispose of SNF in larger packages (e.g., capacity of at least 12 and up to 32 fuel assemblies from pressurized water reactors) would allow similar additional savings in disposal costs. The outlook for deploying multi-purpose canisters in the future depends on progress in disposal planning, and cooperation of the nuclear power utilities, the vendor industry, and government.

1. Introduction: The Canister Quandary

Commercial spent nuclear fuel (SNF) is accumulating in the U.S. at the rate of approximately 2,000 metric tons (MT) per year, from pressurized-water reactors (PWRs) and boiling-water reactors (BWRs). A similar amount of SNF is being transferred each year from nuclear power plant (NPP) fuel pools to dry storage. Most of the welded, sealed stainless steel canisters used for dry storage can also be used for future transport to a centralized storage facility or a geologic repository, hence these are called dual-purpose canisters (DPCs).

By the year 2035 approximately half of all SNF in the U.S. is projected to be stored in DPCs (Figure 1). If nothing changes by about 2060, essentially all the SNF from existing reactors will be in DPCs (nearly 140,000 MT). Unless these DPCs can be disposed of directly, they will all need to be cut open and the SNF repackaged in new containers, at great expense. The root problem is that the DPCs in use, and those presently contemplated for future use, have not been designed or licensed for SNF disposal. This chapter identifies a range of canister design options that could be disposable, and recognizes certain solutions that are being pursued by other countries.

The disposability of SNF depends on effective safety strategies that ensure waste will be isolated in the repository, with sufficient heat dissipation, and control of nuclear criticality. Such strategies depend on repository site characteristics which may vary significantly for different host geologic media. The U.S. does not presently have any specific site in active consideration for a repository, so for a canister to be effectively disposable today it needs to be part of a disposal concept that can be demonstrated to function in multiple media.

The framework presented here for understanding SNF canister disposability takes into account alternative safety strategies and different geologic media. It shows what canister design options are available, and it includes possible strategies for direct disposal of existing DPCs, an option that is under study. Functions of the canister and basket (Figure 2), and of the disposal overpack that contains the canister, are identified for a range of Design Alternatives. Example concepts for the canister and basket are proposed for these alternatives.

The potential advantages of storage canisters that are suitable for disposal, or of direct disposal of existing DPCs, include simpler SNF management, lower cost, less secondary waste (e.g., used DPC hulls), and less worker exposure to radiation during canister operations. The quandary is that canister disposability is closely linked with repository siting and selection of a disposal concept, and while siting is probably decades away, SNF continues to accumulate in DPCs of existing design. A universal canister design might help to alleviate this uneconomic situation, but the technical feasibility and design have not been determined, such as the SNF capacity and the materials used for structural and neutron absorbing elements. The quandary is somewhat unique to the U.S. because of the relatively large quantity of commercial SNF projected to accumulate, and the high reliance on dry storage in sealed, welded canisters.

This is a technical document that does not take into account the contractual limitations under the Standard Contract. Under the provisions of the Standard Contract, DOE (the U.S. Department of Energy) does not consider spent fuel in canisters to be an acceptable waste form, absent a mutually agreed to contract modification.

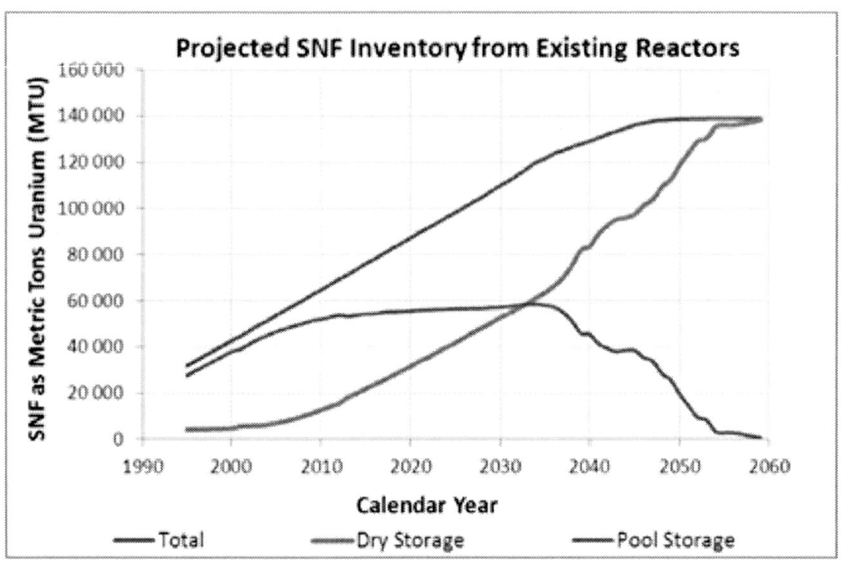

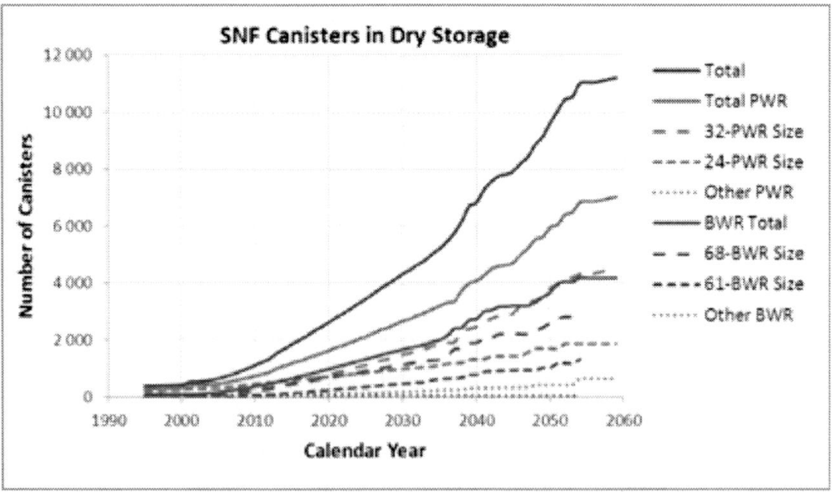

Figure 1. Projected inventory of SNF (upper), and projected number of dry-storage canisters (lower) (from Hardin et al. 2013b).

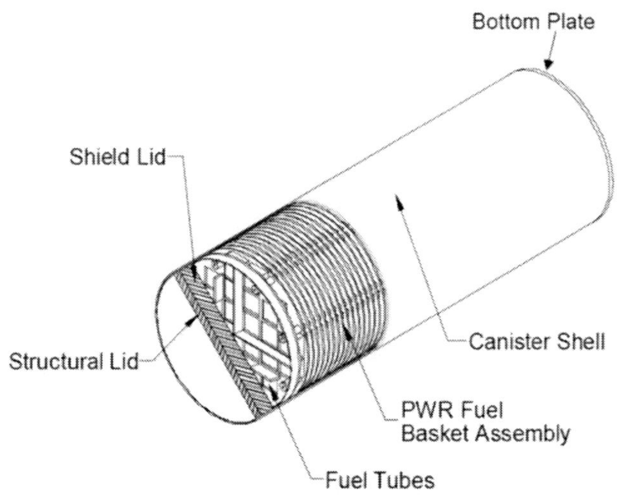

Figure 2. Typical dry storage canister (NAC International TSC-24 shown) for 24 PWR fuel assemblies.

2. BACKGROUND

DPC design has changed since they came into commercial use about 20 years ago: they are larger and use various means to control criticality (see Greene et al. 2013 for a summary of DPC characteristics). Design changes have occurred in parallel with advances in the methods used to analyze thermal and criticality responses during storage and transportation. Disposal has not been a factor in DPC design due partly to the terms of the current contracts between utility companies and the government. The U.S. Department of Energy (DOE) has responsibility for final disposition of the SNF, and the utilities are to deliver it to the DOE as uncanistered ("bare") fuel.

A canister design that could be used by the utilities for storage, and is suitable for disposal also, has been proposed on two previous occasions: the multi-purpose canister (MPC) initiative of the 1990's (DOE 1994), and the 2008 transport-aging-disposal (TAD) canister concept developed for the proposed repository in volcanic tuff (DOE 2008). Both concepts were proposed for a specific site (Yucca Mountain) and both were abandoned. At present, SNF continues to accumulate in DPCs which are not purpose-designed or licensed for disposal.

So how do we start to design canisters that are suitable for SNF disposal? The answer is to look at the physical and chemical conditions of disposal, considering different geologic media. Disposal environments and applicable repository concepts have been investigated for several years in the U.S. Department of Energy's Used Fuel Disposition (UFD) R&D program (Hardin et al. 2011; 2012a; 2013a) as well as internationally. The following sections provide background discussion of disposal concepts, waste isolation strategy, approach to retrievability, thermal management, and nuclear criticality analysis.

2.1. Summary of Alternative Disposal Concepts

A range of disposal concepts needs to be defined in order to delineate how SNF canister Design Alternatives could work. A geologic disposal concept is defined to include a waste form, geologic setting, and engineering concept of operations. For this discussion the waste form is commercial SNF assemblies. The generic (non-site specific) geologic settings being considered by the UFD R&D program include:

- Crystalline, including igneous intrusive or extrusive, and metamorphic rock types
- Argillaceous (clay-rich) sedimentary rock
- Salt (bedded or domal)

The engineering concept of operations includes the waste packaging, emplacement mode (e.g., in-drift, borehole, vertical, horizontal, etc.), and other engineered barriers (e.g., buffer, backfill, plugs, seals, etc.). This chapter describes canister Design Alternatives that could ensure safety of mined geologic disposal in conjunction with other engineered barriers (disposal overpacks, backfill, etc.) and the host rock. Only mined geologic disposal concepts are considered (as opposed to other concepts such as disposal in deep boreholes).

This discussion considers canister Design Alternatives that could possibly replace existing DPC designs in the future, including smaller canisters. For example, canisters containing four fuel assemblies from pressurized-water reactors (4-PWR) are considered in addition to the current 24-PWR, 32-PWR and 37-PWR DPCs. The discussion also includes alternatives that would reopen existing or future loaded canisters for modification, and then reseal them for disposal.

2.1.1. Enclosed Emplacement Mode Concepts

An important distinction in disposal concepts is whether waste packages are emplaced in direct contact with a surrounding medium such as buffer, backfill, or host rock. This is contrasted with open modes in which packages are emplaced with surrounding, connected air spaces that can be ventilated to remove heat. The following enclosed modes are based on international experience and previous repository concept studies in the U.S. (Mariner et al. 2012; Hansen and Leigh 2011; Hansen et al. 2010).

Crystalline Rock, Enclosed (Clay Buffer) Concept – A repository constructed at several hundred meters depth in crystalline rock (igneous or metamorphic). Corrosion resistant waste packages would be installed in large-diameter borings, surrounded by buffer material consisting of swelling clay that has been dehydrated and compacted into blocks. The buffer would be installed at emplacement (e.g., Swedish KBS-3 concept; SKB 2011, Section 5.5). Possible corrosion resistant materials for the waste package outer layer include copper and titanium (SKB 2011, Section 5.4; Shoesmith et al. 1995). Waste packages would be relatively small with limited heat output to maintain buffer temperature less than 100°C (e.g., 4-PWR size with less than 1,700 W heat output at emplacement; SKB 2011, Sections 5.2 and 5.3). For the Swedish KBS-3 concept the fuel basket is proposed to take the form of a cast iron insert that separates the fuel assemblies (but contributes substantial weight). Access and service tunnels (also called drifts) would be backfilled with low-permeability, clay-based backfill at closure (Hardin et al. 2012a, Section 1.4.5.1; Hardin et al. 2013a, Section 4.1).

Salt Concept – A repository constructed at approximately 500 to 1,000 meters depth in bedded or domal salt. Disposal overpacks would consist of thick carbon or low-alloy steel, and waste packages would be emplaced on the floor in drifts or alcoves, and immediately backfilled with crushed salt (e.g., from excavating the next emplacement drift). This concept is similar to an option developed in Germany (Graf et al. 2012) and to a concept developed for heat-generating high-level waste glass (Carter et al. 2011). Waste packages of any size (including 32-PWR size or larger) could be used with heat output limited to approximately 10 kW at emplacement (Hardin et al. 2013a, Section 4.2). The fuel basket would be designed to meet preclosure structural and criticality control requirements. Any liquid water present in the repository would be chloride brine. All repository openings would be backfilled at closure (Hardin et al. 2012a, Section 1.4.5.2; Hardin et al. 2013a, Section 4.2).

Clay/Shale Enclosed Concept – A repository constructed at several hundred meters depth in clay-rich, low-permeability sedimentary rock. Waste packages would be emplaced in steel-lined horizontal borings, surrounded by clay-based buffer material. Waste packages for SNF could be made from carbon steel or other corrosion allowance material. This concept is similar to that developed in France for SNF waste (Andra 2005, Section 4.5.2). Waste packages would be small with limited heat output (e.g., 4-PWR size with heat output on the order of 1 kW at emplacement; Hardin et al. 2012a, Section 3.1.2). The fuel basket (including neutron absorber components, if any) would be designed to meet preclosure structural and criticality control requirements. Access drifts would be backfilled with low permeability clay-based backfill at closure (Hardin et al. 2012a, Section 1.4.5.3; Hardin et al. 2013a, Section 4.3).

2.1.2. Open Emplacement Mode Concepts

Open modes can be used to achieve thermal goals with larger packages containing more than four PWR assemblies, or equivalent SNF from boiling water reactors, in host media other than salt (for which waste packages could be backfilled immediately). Waste packages would be placed horizontally on the emplacement drift floor, either parallel or transverse to the axis of the drift. Emplacement drifts would remain open for cooling, remote inspection, and maintenance for as long as 50 years. Drift opening stability could be important for underground design, especially in soft sedimentary rock. Long-term stability of drift openings is one reason for limiting the operational period to 50 years (Hardin et al. 2012a, Section 1.5) but longer operations with little or no maintenance might be achieved.

Hard Rock Unsaturated, Unbackfilled, In-Drift Concept – A repository constructed and operated in competent, hard rock (e.g., igneous or metamorphic) using in-drift emplacement and forced ventilation for up to 50 years. Disposal overpacks would be made from materials that resist corrosion in chemically oxidizing conditions (e.g., BSC 2007a; DOE 2006). The hydrologic setting would be unsaturated, so backfill would not be needed but other engineered barriers might be installed such as long-lived shields to divert downward water percolation (DOE 2008; Hardin et al. 2012a, Section 1.5.3; Hardin et al. 2013a, Section 4.6.1).

Hard Rock Saturated, Backfilled, In-Drift Concept – A repository constructed and operated in competent, hard rock in a saturated hydrologic setting. This concept would use in-drift emplacement with forced ventilation for up to 50 years. Disposal overpacks would be designed to perform in the disposal environment (i.e., oxidizing or reducing conditions, or both). A low permeability buffer or backfill would be installed prior to closure, to condition the chemical and physical environment at the waste package surface, and to limit groundwater flow along repository openings. Backfill would be installed remotely, or directly if waste packages are self-shielding. Unlike argillaceous media discussed above, hard rock typically has numerous, permeable fractures. Thus, a principal function of backfill would be to impede groundwater flow, so low-permeability materials would be required (Hardin et al. 2013a, Sections 4.6.2 and 6.3.2). Postclosure performance would be similar to the crystalline enclosed concept discussed above. This backfilled concept could also be implemented in unsaturated formations (Hardin and Sassani 2010).

Argillaceous Rock, Backfilled, In-Drift Concept – A repository constructed and operated in soft, clay-rich sedimentary rock, with in-drift emplacement, and forced ventilation for up to 50 years after emplacement. Emplacement, access, and service drifts would be backfilled at closure, with a low-permeability engineered material (Hardin et al. 2012a, Section 1.5.2; Hardin et al. 2013a, Section 4.5). Backfill would be installed either remotely, or directly if waste packages are self-shielding. Options for backfill materials are discussed by Hardin and Voegele (2013, Appendix B). Backfill functions would include low permeability, and mechanical support after roof collapse so as to limit the extent of damage in the host formation. Disposal overpacks could include an outer layer of corrosion resistant material, however, less resistant packaging could be used if a complementary containment function were demonstrated for the backfill (Hardin et al. 2013a, Sections 4.5 and 6.3.3). Postclosure performance would be similar to a reference concept for Opalinus clay that uses in-drift emplacement and clay-based backfill (NAGRA 2002; 2003; 2009).

2.2. Waste Isolation Performance

A geologic repository is a system, with waste isolation functions typically described as: 1) limiting water that contacts waste forms; 2) limiting rates of radionuclide release; and 3) attenuating radionuclide concentrations along potential transport pathways. The first of these (water contact) would be assigned to the disposal overpack, while the second (rates of release) would be assigned to both the overpack and the waste form. The third function (attenuated transport) would be assigned to engineered and natural barriers outside the canister. The canister and fuel basket would not necessarily contribute to any of these functions. This is consistent with precedent (DOE 2008; 2006) wherein no long-term containment credit was taken for the SNF canister. It is also consistent with expected limitations on the corrosion lifetime of existing DPCs in disposal environments. Note that in addition to containment performance, the overpack would also provide the mechanical support needed for waste transport underground and final emplacement.

Depending on the disposal concept, the overpack could consist of a single layer or multiple layers of different materials. There are significant differences in the behavior of candidate overpack materials, especially with respect to corrosion processes. Numerous analyses have been developed to describe corrosion behaviors (BSC 2008, for example, Sections 1.1.07.00.0A, 2.1.03.10.0A and 2.1.09.03.0B). Corrosion allowance materials (e.g., cast iron, carbon steel, or low-alloy steel) and corrosion resistant materials (copper, titanium, Ni-Cr-Mo alloys) are mentioned frequently in Section 4 below. However, the primary focus of this chapter is not the overpack, but the design and performance of existing and future canisters. Assigning containment and structural functions solely to the overpack simplifies the treatment of canisters, and while overpack concepts are discussed, details are not elaborated.

Compliance with U.S. Nuclear Regulatory Commission (NRC) postclosure performance requirements (e.g., 10CFR63 Subpart L) can be assumed for all the alternative design concepts considered here (with appropriate performance of overpacks and other engineered and natural barriers). Compliance would actually need to be demonstrated through performance assessment analysis and supporting process models (for example, see DOE 2008). Waste isolation is best achieved through redundant natural and engineered barriers including the host rock, disposal overpack, and the rate-limited dissolution of SNF. Note that for some of the alternative canister/basket concepts discussed in Section 4, overpack containment integrity is used to exclude groundwater, or reduce the probability for groundwater flooding, for control of postclosure criticality.

2.3. Waste Package Retrievability

Retrievability requirements are covered in the NRC regulations on geologic disposal (see 10CFR60.111(b) and 10CFR63.111(e)), and are generally posed to ensure removal of emplaced waste from a repository should there be a safety concern of sufficient magnitude to warrant such action. For open emplacement modes, retrieval prior to backfilling would involve picking up packages and transporting them to the surface by the same means used to emplace them. Handling and containment functions would be assigned to the disposal overpack (the same functions assigned for emplacement). In salt or any medium in which backfill is emplaced soon after emplacement, or in the event of rockfall or roof collapse, retrieval would involve excavation. In such cases any additional containment functions would also be assigned to the disposal overpack. Accordingly, SNF canisters need have no retrievability function.

2.4. Thermal Management

For thermal analysis discussed here, decay storage was limited to 100 years, and repository or panel operation after waste emplacement (such as ventilation, prior to panel closure) was limited to 50 years. The overall objective to limit storage and repository operations to 150 years is comparable to previous plans (DOE 2008) which would have closed a repository when the oldest SNF would have been approximately 150 years out-of-reactor. The 100-year decay storage limit is also similar to the 60-year and 160-year timeframes considered in the NRC's draft "waste confidence" environmental impact statement (NRC 2013). These limits imply the capability for safe storage and transport of SNF to the repository for 100 years or longer after reactor discharge. They also imply that underground openings can remain stable with little or no maintenance for 50 years.

The terms repository closure and panel closure are used interchangeably in this discussion. The concept of panel closure is needed because commercial SNF will be generated in the U.S. over about 90 years (from 1965 to 2055, considering presently operating reactors and no new builds; Carter et al. 2012). A geologic repository could therefore operate for much longer than 50 years if a similar disposition path is used for all the SNF. However, disposal areas within a repository, designated as panels, could be operated and closed periodically.

Final disposal of SNF will not necessarily take until calendar 2205 (final plant shutdown in 2055 plus 150 years) because in the coming decades there will be a point in time when the repository site and disposal concept are known, and from that point forward canisters and other packaging can be designed for earlier disposal. Therefore, capability for earlier disposal could be an important factor in choosing the design for future, disposable SNF canisters.

Thermal analyses discussed in this chapter are based on limiting waste package thermal power at repository closure, to control the peak postclosure temperature in the host rock or at the waste package surface. Peak temperature limits can control threshold responses such as groundwater boiling, other phase changes in the host rock or backfill, and decrepitation of the host rock. Peak temperature at the waste package surface correlates closely with instantaneous thermal power, and package diameter is a second-order effect (Hardin et al. 2012b, Section 5). Other types of thermal constraints, such as time-temperature integrated measures, may also apply (e.g., see BSC 2008a, Section 6.5). Integrated long-term power is generally greater for larger packages than smaller ones, even if aging is used to limit peak power at closure, because of greater inventory of heat-generating nuclides with intermediate half-lives (e.g., ^{241}Am is not significantly reduced by decades of decay storage).

It is possible to perform useful thermal analyses using only thermal conduction physics, plus radiative heat transfer if there are any opening spaces around the waste packages after repository closure. Convective heat transfer can occur in porous media such as backfill or rock debris, or in fractures, but is limited especially in unsaturated settings (Stauffer et al. 1997). Also, host media and backfill materials are typically selected for low permeability in addition to other attributes.

Thermal conductivity of the host geologic medium is a key parameter, and can be characterized as low, medium, or high (less than 2, 2 to 3, and greater than 3 W/m-K, respectively). Average thermal conductivity for argillaceous media is typically low, while crystalline rock is medium and salt is high (Hardin et al. 2012b). Materials used for buffer and backfill are assumed to be dry at emplacement and throughout the maximum thermal period (i.e., up to a few hundred years). This is appropriate because maximum temperatures in clay-based engineered materials could exceed boiling, in which case they would resist hydration. Rehydration timing is uncertain and could take hundreds of years with low permeability in the host rock and low permeability in partially rehydrated engineered barriers.

Temperature tolerances of the host rock and engineered barrier materials also have an important impact on thermal management. For salt and hard rock settings, a host rock peak temperature of 200°C is reasonable based on many years of investigations (Hardin et al. 2012a; 2013a). For argillaceous media a host rock peak temperature target of 100°C is similar to limits imposed by international programs (SKB 2011, Section 5.2.1; Andra 2005, Section 1.2.3.4). For backfill materials, higher limits are needed (of the order of 150°C or greater) for disposal of larger waste packages (greater than 4-PWR size) in the 150-year timeframe discussed above. Thermal limits for argillaceous host media and clay-based backfill materials are the focus of active investigation in the UFD R&D program.

An exception to the use of simple thermal models arises when material properties change significantly during the thermal period. For example, for the salt repository calculations discussed below a numerical finite-element method is used primarily to represent backfill consolidation which increases thermal conductivity by an order of magnitude (Hardin et al. 2012a, Appendix C).

Repository maximum temperature behavior is determined by fuel burnup, fuel age at closure, waste package capacity, repository geometry including spacings, and properties of the host rock and engineered materials. Fuel burnup refers to the amount of energy produced from fission in power plants, which in turn determines the heat output of SNF for disposal. Analyses have shown that repositories in salt, hard rock and argillaceous media could meet peak temperature targets, with a range of fuel burnup, and repository panel closure when the SNF age is 70 to 150 years (Hardin et al. 2012a; 2013a). The following paragraphs summarize thermal analysis results for different disposal concepts:

Enclosed emplacement modes (crystalline and clay/shale media) – Thermal results for crystalline and clay/shale enclosed concepts are similar because of the similarity of the clay-based buffer and clay/shale host media. Where used, the clay-based buffer constitutes the dominant thermal resistance. SNF with high burnup (up to 60 GW-d/MT) could be emplaced in 4-PWR waste packages after approximately 100 years of surface decay storage, without exceeding the 100°C peak temperature target for clay-based materials in this concept. This result is similar to SNF management practices being implemented by the Swedish program. Waste packages containing a single high-burnup SNF assembly could be emplaced after approximately 10 years of surface decay storage.

Salt enclosed concept – High thermal conductivity of intact rock salt, and tolerance for elevated temperatures (200°C), allow waste packages to be emplaced relatively early and the repository to be backfilled immediately. Thermal management for DPC direct disposal in salt is illustrated with a reasonably bounding calculation, with 32-PWR size packages containing SNF with 60 GW-d/MT burnup, emplaced and backfilled at 70 years after reactor discharge (Figure 3). The calculation shows that a 200°C salt peak temperature limit can be met using a simple disposal concept (Figure 4). Such calculations are performed using temperature-dependent rock properties, and full thermal-mechanical coupling to represent thermally activated creep of the intact salt and crushed salt backfill (Hardin et al. 2013a).

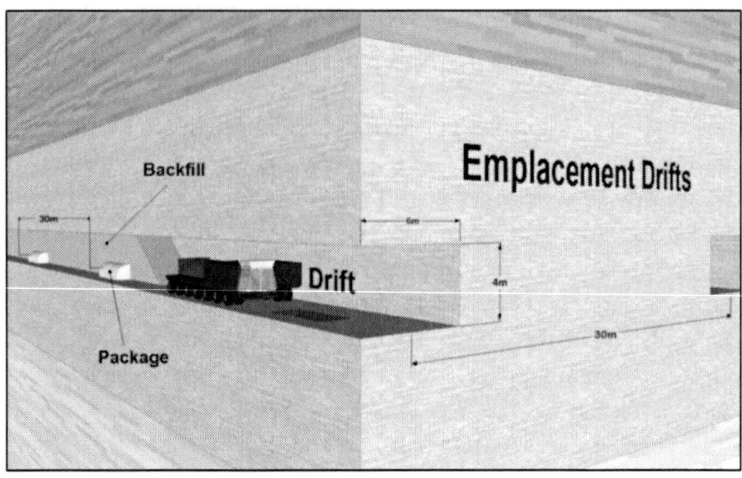

Figure 3. Schematic concept for disposal of large waste packages (e.g., containing DPCs or other large SNF canisters) in a salt repository.

Open emplacement modes – Open, ventilated emplacement is generally required for non-salt media (hard rock and argillaceous) unless waste package capacity is approximately 4-PWR size or smaller, or unless surface decay storage can extend for hundreds of years (Hardin et al. 2012a). This statement results from thermal analysis of enclosed emplacement modes such as those being investigated in European disposal R&D programs, and temperature limits for clay-based buffer and backfill materials (e.g., 100°C). Note that open modes permit extended repository ventilation, but the emplacement drifts may still be backfilled at or before repository closure as discussed below.

Thermal histories for open emplacement modes with large waste packages (21-PWR size) were evaluated by Hardin et al. (2012a, Section 3.2). Direct disposal of typical existing DPCs (32-PWR size) using open modes was also evaluated (Hardin et al. 2013a, Section 5). The results from these calculations are summarized as follows:

- The peak temperature target (200°C) for crystalline rock (thermal conductivity 2.5 W/m-K) at the emplacement drift wall can be readily met for all package sizes, up to 60 GW-d/MT burnup, within the 150-year assumed disposal timeframe.
- For argillaceous media (1.75 W/m-K) the host rock peak temperature target (100°C) cannot be met for moderate to high burnup, without: 1) increasing drift and package spacings; and/or 2) increasing the

timeframe (e.g., 200 years). Alternatively, a region of host rock around each waste package of approximately 1-meter thickness, could be heated to a higher temperature.

- Installation of backfill at or before closure dramatically increases peak temperature at the waste package surface, such that backfill peak temperature tolerance of 150 to 200°C is needed for waste packages larger than 4-PWR size, within the 150-year timeframe.

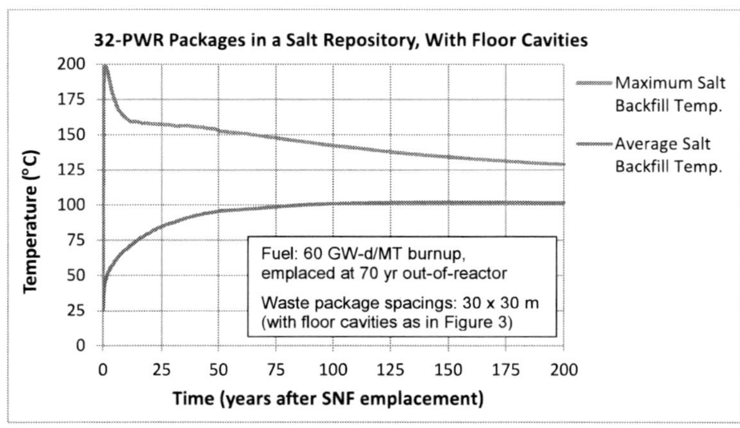

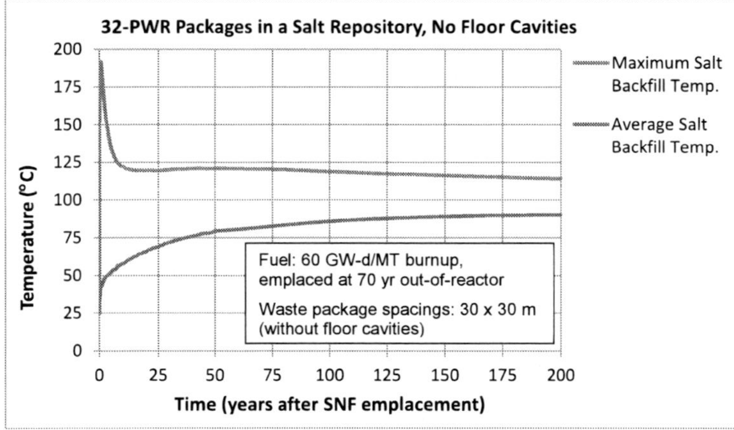

Source: Hardin et al. (2013a, Figure 5-5). Note: Floor cavities are semi-cylindrical cavities cut to the diameter of waste packages, into which packages would be emplaced to facilitate heat transfer to the host salt.

Figure 4. Temperature histories for high burnup SNF in 32-PWR size packages, for the salt concept, showing one way to lower temperature with in-drift emplacement: floor cavities (upper) compared to emplacement directly on the floor (lower).

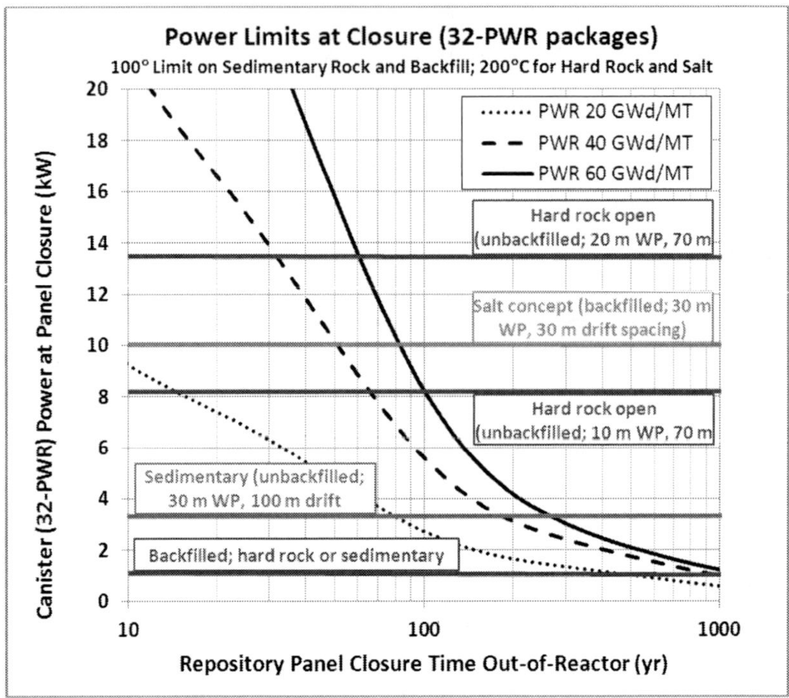

Figure 5. Summary of disposal timing for SNF with different fuel burnup (20, 40, and 60 GW-d/MT), in 32-PWR size packages, for various disposal concepts (salt, hard rock, and sedimentary, unbackfilled and backfilled). Curves represent thermal decay for 32 PWR assemblies, and intersections with horizontal lines show minimum cooling time before repository closure (after Hardin et al. 2013c).

The timing of open-mode disposal of SNF in large (32-PWR size) waste packages is summarized in Figure 5. For smaller waste packages (e.g., 21-PWR size), smaller repository spacings and earlier disposal are possible (Hardin et al. 2012a, Section 3.2) but if backfill is installed the peak temperature will exceed 100°C except for smaller waste packages (4-PWR size or smaller).

Another thermal management tool that is available, especially with loading of large canisters, is thermal de-rating of the canisters by loading fewer SNF assemblies (e.g., as few as 24 assemblies in a canister designed for 32). For example, de-rating might be used with disposal of a relatively small portion of high-burnup SNF in large canisters, in argillaceous media, to meet the host rock peak temperature target. Alternatively, it might arise for backfilled concepts in order to keep the peak waste package surface temperature below a limit associated with protecting the SNF cladding from temperatures greater than 350°C (NRC 2003). Both of these de-rating situations might occur only for the youngest SNF and/or SNF with the highest burnup, so it might be inappropriate to let them dictate major features of the repository design such as the schedule for closure. Instead, the repository could be designed for low and moderate burnup, and a de-rating strategy could be used for high-burnup SNF whereby fewer assemblies are loaded into individual canisters.

2.5. Postclosure Criticality Control

SNF canisters that could be considered for geologic disposal will be subcritical in the disposal environment unless they are: 1) flooded with groundwater; and 2) the canister internals (e.g., basket and neutron absorbers) are degraded by corrosion. For transportation licensing (10CFR71.55) canisters are typically analyzed in a flooded condition, within a transportation overpack. For most of the canisters licensed for transport, neutron absorber features such as Boral® sheets are installed at manufacture and prevent criticality under such temporary accident conditions. For disposal criticality a similar flooded analysis may also be used, but with basket and neutron absorber components degraded from long-term exposure to groundwater. Analysis to-date suggests that some existing canisters may have conservative design, and/or may be loaded with less reactive SNF, so that degradation of the internals could occur without causing criticality (Hardin et al. 2013a; Sections 8 and 10). Direct disposal of such canisters is discussed in Section 4.3 for Design Alternative III.D. The following discussion focuses on the significant fraction of DPCs for which flooding and degradation of internals would likely lead to criticality.

Neutron absorbing features used in existing SNF canisters have been summarized (Hardin et al. 2013a, Section 3.1.4) and the materials used in existing canisters have been tabulated (Greene et al. 2013). Features include absorber plates interposed between adjacent assemblies, flux traps, control rods, depleted burnable poisons in different configurations, etc. Absorbing materials used in different canister designs typically include the aluminum-boron-carbide cermet Boral®, metal matrix composites, Metamic®, Bortec®, Tetrabor®, and Carborundum B_4C (Rigby 2010).

Analysis of postclosure criticality in accordance with 10CFR Part 63 must consider whether the total probability is greater than 10^{-4} for 10,000 years after disposal (i.e., 10^{-4} per repository realization, see 10CFR63.102(j)). If the probability is greater than 10^{-4} then a consequence assessment may be performed, and the result combined probabilistically (i.e., as risk) with the dose consequences from other repository release pathways (DOE 2003, Section 3.7). This chapter emphasizes strategies for limiting the total probability of postclosure criticality to less than 10^{-4}. Even if a consequence approach were selected, one or more of the alternatives discussed in Section 4 would still be used to limit the probability of a criticality event.

The timeframe for criticality analysis is assumed to be 10,000 years based on prior regulation (10CFR63.114). Dose assessment is extended beyond 10,000 years, through the "period of geologic stability." However, for screening criticality FEPs under the assumed regulatory framework (see Section 3 below) only a 10,000-year regulatory period needs to be analyzed.

As stated above, flooding and degradation of fuel canisters are necessary for criticality. If the geologic setting or disposal overpack excludes groundwater for at least 10,000 years, then the SNF will remain subcritical for that time. Some host geologic media may exclude groundwater over the long term, such as the salt concept. In host media where groundwater is abundant, or may become so in the future, overpack containment performance by itself may be insufficient for excluding SNF criticality on the basis of low probability (less than 10^{-4} per repository realization). That is because studies of manufacturing defects show that the probability of an early package failure (generalized to result in waste package breach) is of the order of 10^{-5} per package (BSC 2007b, Table 7-1) whereas a value less than 10^{-8} per package (i.e., per disposal overpack) would be needed to exclude criticality for 10,000 waste packages for which flooding results directly from breach. The possibility of such a high-reliability containment is addressed by Design Alternative III.A. External events such as seismicity and faulting might also contribute to overpack breach, increasing the total probability of criticality.

Once canisters are flooded, criticality control depends on the configuration of the degraded internals. Degraded configurations for an open, hard rock, unsaturated disposal concept have been analyzed extensively (BSC 2008c; 2008d). In more recent numerical simulations (Clarity and Scaglione 2013) two degraded configuration cases were considered: 1) a loss-of-absorber case in which neutron absorber plates are replaced by groundwater; and 2) a basket degradation case in which both the absorber plates and the basket are removed and the assemblies are positioned together tightly. Postclosure criticality analyses are reported for existing DPCs from two storage sites, using burnup credit (i.e., neutron interaction with transmuted isotopes) for 28 nuclides, actual assembly burnup data, and actual canister loading data. The results showed that some, but not all, of these DPCs could be subcritical after flooding with groundwater, even with highly degraded internals. These are stylized configurations designed to be reasonably conservative, and additional realism might decrease reactivity. For example, reactivity might be decreased by including in the model: metal corrosion products (displacing water as a neutron moderator), irradiated control rods (displacing water), and fuel degradation (lowering the hydrogen content of interstices between fuel rods). Analysis of degraded configurations for DPC disposal is an area of ongoing investigation (Hardin et al. 2013a, Sections 8 and 10).

Canister misloading is a configuration anomaly that could contribute to nuclear reactivity in the event of waste package breach and flooding. The probability of misloading a SNF assembly into a waste package is of the order of 10^{-5} per assembly, and the joint probability of a misload leading to the potential for criticality is of the order of 10^{-7} (BSC 2008d, Section 4.1.5). These probabilities might be decreased using additional direct measurements for verification of loaded canisters before sealing (i.e., in fuel pools). Criticality analysis numerical methods and supporting data are beyond the scope of this chapter, but current reviews and technical reports are available (Wagner et al. 2011).

3. ASSUMPTIONS

Assumptions have been identified for evaluating the possibility for direct disposal of SNF in existing DPCs (Hardin et al. 2013a; Hardin and Howard 2013), and by analogy, for evaluating other canister alternatives. Highlights include:

- Disposal timeframe is limited to 150 years after discharge (maximum of 100 years decay storage, and 50 years of repository operations such as forced ventilation).
- Regulatory context for disposal licensing will be similar to 10CFR63 and 40CFR197, including a 10,000-year performance period for screening criticality FEPs.
- Reactor operating records can be accessed to provide more realistic inputs to criticality models.
- Criticality consequence analysis may be incorporated into FEP screening arguments and repository performance assessment.

With some exceptions, these assumptions could be applicable to evaluations needed to design and license new canister designs intended for direct disposal.

4. SNF CANISTER/BASKET DESIGN ALTERNATIVES

This section offers a set of alternatives for overpack, canister, and basket characteristics to achieve waste isolation, thermal management goals, and postclosure criticality control. This constitutes a range of approaches for addressing the requirement from 10CFR72:

> (m) To the extent practicable in the design of spent fuel storage casks, consideration should be given to compatibility with removal of the stored spent fuel from a reactor site, transportation, and ultimate disposition by the Department of Energy. §10CFR72.236(m)

The Design Alternatives and examples are summarized in Table 1. Some observations common to all the alternatives are discussed here for brevity:

As noted previously, waste containment and structural functions are assigned to the disposal overpack. Unless disposal overpack containment is specifically discussed for the alternatives, overpacks could be constructed either from corrosion allowance materials (e.g., carbon steel) with containment lifetime of the order of 10^4 years, or from corrosion resistant materials (e.g., Ni-Cr-Mo alloys) with containment lifetime of 10^5 to 10^6 years. The selection would depend on the safety strategy for a particular Design Alternative.

The term "multi-purpose canister" (MPC) is defined to mean a canister that could be licensed for storage, transportation, and disposal. The term "bolted closure" refers to mechanical closures with seals, that can be opened and resealed without cutting the canister or lid. Thus, "bolted" closures could involve threaded lids, swaging, etc. and would not necessarily mean the use of bolts. The term "resealing" is reserved for bolted closures, and "rewelding" for welded canisters that have been cut open for modification of internals, but will be reused. All canister alternatives discussed here would be MPCs, but some would have bolted closures, some welded closures, and some would have combinations. All would include redundant closures (here called inner and outer lids) as required by 10CFR72.236(e).

Table 1. Summary of Design Alternatives for SNF canister disposability

Group	Design Alternative	Example	Description
Group I Small canisters (up to 4 PWR assemblies or BWR equiv.) with no neutron absorbers	I.A Small capacity waste package	I.A.1 Example: Small capacity waste package, welded closure	Canisters closures could be welded without sacrificing flexibility for disposal in a full range of geologic settings, because criticality control may not depend on whether flooding occurs, or on groundwater composition (Section 4.1). This type of canister could also be used in the corresponding can-within-can alternative (Example II.C.1).
		I.A.2 Example: Small capacity waste package, bolted closure	Bolted canisters could preserve flexibility to modify the internals (e.g., add fillers; Section 4.5), or unload the SNF and reuse the canister. This type of canister could also be used in the corresponding can-within-can alternative (Example II.C.2).
	I.B Small capacity waste package combined with heavy insert	I.B.1 Example: Small capacity waste package combined with heavy insert, welded closure	Canisters closures could be welded without sacrificing flexibility for disposal in a full range of geologic settings, because criticality control would not depend on whether flooding occurred, or on groundwater composition.
Group II Larger canisters with bolted closures or bolted/welded combinations	II.A Bolted closures (both inner and outer canister lids)	II.A.1 Example: Larger canisters, bolted inner and outer closures	Bolted canisters could enhance flexibility to reopen and add reactivity control improvements or unload the SNF and reuse the canister. Postclosure performance of reactivity control improvements (e.g., "surrogate" control rods, Section 4.2) might depend on congruent degradation with SNF rods. Canister design could facilitate later installation of a third, welded lid as needed for transportation or disposal.
	II.B Bolted closure combined with welded closure	II.B.1 Example: Larger canisters, bolted inner and welded outer closures	The outer lid could be cut off and removed (e.g., dry) for access to the inner bolted lid without exposing fuel. The inner lid and shield plug could be readily removed for modification of absorber and structural components prior to disposal, then replaced and resealed. Canister design could add the capability to re-weld another outer lid, even if the original outer lid is cut off by cutting through the canister wall.

Table 1. (Continued)

Group	Design Alternative	Example	Description
		II.B.2 Example: Larger canisters, welded inner and bolted outer closures	Removing the outer bolted lid would be relatively straightforward and could be done dry, but cutting out the inner lid would involve essentially the same choices as for existing (welded) DPC designs discussed in Group III.
	II.C Can-within-can arrangements using bolted or welded closures	II.C.1 Example: Larger can-within-can systems with welded inner/welded or bolted outer canisters	Inner canisters would have welded closures (similar to Example I.A.1), and could be cool and nonreactive enough to never require reopening for disposal. The outer canister lid could be welded or bolted. However, the inner canisters could be credited for storage and transportation licensing as containment envelopes for moderator exclusion, reducing performance requirements on a bolted outer canister.
		II.C.2 Example: Larger can-within-can systems with bolted inner/welded or bolted outer canisters	Inner canisters would have bolted closures (similar to Example I.A.2). These could also be cool and non-reactive enough to never require reopening for disposal, and bolted closures may have operational advantages during loading and dewatering. The outer canister lid could be welded or bolted, but a welded lid could simplify monitoring of canister integrity during storage.
Group III Existing DPC designs (welded closures) used for disposal without reopening	III.A High reliability (10^{-8}/yr) overpack performance	III.A.1 Example: Existing DPC designs with high-reliability overpacks	High-reliability disposal overpack performance prevents breach and DPC flooding and degradation, in a saturated or unsaturated environment, for at least 10,000 years.
	III.B Disposal in unsaturated conditions with multiple engineered and natural groundwater exclusion/diversion barriers	III.B.1 Example: Existing DPC designs loaded with relatively non-reactive SNF	Canisters remain subcritical even after flooding with groundwater of dilute composition, and degradation of basket structure and neutron absorbers. Demonstrated using analysis of degraded configurations, with as-loaded assembly arrangements and fuel burnup characteristics.

Group	Design Alternative	Example	Description
		III.B.2 Example: Existing DPC designs protected by multiple barriers	Multiple natural and redundant engineered barriers prevent DPC flooding (e.g., in an unsaturated disposal environment) for at least 10,000 years.
	III.C Flooding possible only with chloride brine	III.C.1 Example: Existing DPC designs in a salt repository	Disposal of DPCs in a salt repository, with thick-walled steel overpacks, prevents flooding and mitigates the consequences of any condition that might cause flooding (i.e., with chloride brine).
	III.D Disposal of existing DPCs containing relatively non-reactive SNF, so that packages are subcritical after flooding and degradation of canister internals	III.D.1 Example: Existing DPC designs loaded with relatively non-reactive SNF	Canisters remain subcritical even after flooding with groundwater of dilute composition, and degradation of basket structure and neutron absorbers. Demonstrated using analysis of degraded configurations, with as-loaded assembly arrangements and fuel burnup characteristics.
Group IV Existing DPC designs (welded closure) used for disposal with modifications	IV.A Reactivity control improvements	IV.A.1 Example: Reactivity control improvements in DPCs at initial loading	Future disposal of existing DPC designs, with reactivity control improvements installed at initial loading, for postclosure criticality control.
		IV.A.2 Example: Reactivity control improvements in reopened/rewelded DPCs	Future disposal of existing DPC designs, with reactivity control improvements installed after reopening. Canisters would be cut open, modified, and rewelded.
	IV.B Fillers for reactivity control	IV.B.1 Example: Fillers installed in DPCs at initial loading	Future disposal of existing DPC designs, with filler installed at initial loading, for postclosure criticality control.
		IV.B.2 Example: Fillers installed in reopened/rewelded DPCs	Future disposal of existing DPC designs, with filler installed after reopening (e.g., via dewatering ports). If canisters are cut open for filler installation (i.e., lids removed), they would be reused, by rewelding lids or by welding on a new head that fits on the canister end.

Table 1. (Continued)

Group	Design Alternative	Example	Description
Group V Multi-purpose canisters with long-lived baskets and neutron absorbers	V.A Larger canisters, with basket and absorber materials selected for longevity in oxidizing environments	V.A.1 Example: Larger canisters and baskets designed for ≥10,000-year life exposed to oxidizing disposal environments	Larger canisters with stainless steel based canister shell and basket materials, designed to maintain initial SNF basket configuration and neutron absorption for >10,000 years exposure to groundwater or humidity environments, in oxidizing conditions.
	V.B Larger canisters, with basket and absorber materials selected for longevity in a range of chemically reducing and oxidizing environments	V.B.1 Example: Larger canisters and baskets designed for ≥10,000-year life exposed to reducing or oxidizing disposal environments	Larger canisters with corrosion resistant canister shell and basket materials, designed to maintain initial SNF basket configuration and neutron absorption for >10,000 years exposure to groundwater or humidity environments, in reducing or oxidizing conditions.

Table 2. Applicability of SNF canister disposability Design Alternatives to disposal concepts

	Disposal Concepts >>>>	Enclosed		Open		
		Crystalline	Clay/Shale	Salt	Hard Rock	Argillaceous
Group I – Small canisters (up to 4 PWR assemblies or BWR equivalent) with no neutron absorbers	I.A Small capacity waste package	√	√	a	a	a
	I.B Small capacity waste package combined with heavy insert	√	√	a	a	a
Group II – Larger canisters with bolted closures or bolted/welded combinations	II.A Bolted closures (both inner and outer canister lids)	b	b	√	√	√
	II.B Bolted closure combined with welded closure	b	b	√	√	√
	II.C Can-within-can arrangements using bolted or welded closures	√ (b)	√ (b)	√	√	√
Group III - Existing DPC designs (welded closures) used for disposal without reopening	III.A High reliability (10^{-12}/yr) overpack performance	b	b	c	√	√
	III.B Disposal in unsat. conditions with multiple natural and engineered groundwater diversion/exclusion barriers	b	b	d	√	d
	III.C Flooding possible only with chloride brine	e	e	√	e	e
	III.D Disposal of existing DPCs so that packages are subcritical after flooding with dilute ground water, and degradation of canister internals	b	b	√	√	√
Group IV - Existing DPC designs (welded closure) used for disposal with modifications	IV.A Reactivity control improvements	b	b	c	√	√
	IV.B Fillers for reactivity control	b	b	c	√	√

Table 2. (Continued)

		Enclosed		Open		
		Crystalline	Clay/Shale	Salt	Hard Rock	Argillaceous
Group V – Multi-purpose canisters with long-lived baskets and neutron absorbers	V.A Larger canisters, with basket and absorber materials selected for longevity in oxidizing environments	b	b	c	√	f
	V.B Larger canisters, with basket and absorber materials selected for longevity in a range of chemically reducing and oxidizing environments	b	b	c	√	√

Notes:

√ Directly applicable.

Shaded cells represent combinations that would not be appropriate based on cost or technical considerations.

a) Small packages (up to 4-PWR size) could be workable for these disposal concepts, but are unnecessary.

b) Larger canisters (>4-PWR size, and the outer canister for the can-within-can) could require too much decay storage (>>150 years) to meet peak temperature targets.

c) These a would add confidence in performance of a salt repository, but are likely unnecessary.

d) Host media for these disposal concepts are nominally saturated.

e) Chloride brine of sufficient concentration would likely be found only in evaporite deposits (especially a salt repository)

f) Argillaceous media typically have low permeability and organic content that render them chemically reducing.

4.1. Group I – Small Canisters (Up to 4 PWR Assemblies or BWR Equivalent) With No Neutron Absorbers

Small capacity waste packages can be used with both enclosed emplacement modes (Section 2.1.1) or open modes (Section 2.1.2), and therefore offer maximum flexibility for thermal management in disposal. The alternatives in this group differ with respect to the type of basket used to position the SNF (fabricated from plate and tube, or solid cast insert). Bolted or welded closures may be used.

Design Alternative I.A – Small capacity waste package.

Capacity: Up to 4 PWR assemblies (or BWR equivalent) depending on enrichment and burnup.

Overpack: Mechanically robust for transport and handling. Provides containment for waste isolation, for a duration that is consistent with the safety strategy.

Canister and Basket: Maintains positions of assemblies after flooding with groundwater (to be consistent with analyzed configurations) for at least 10,000 years after waste package initial breach (which might occur soon after repository closure). Maintains material compatibility with a range of potential overpack designs.

Discussion: If canister capacity is small enough, criticality control for intact or degraded canisters flooded with groundwater could require no internal absorbers. Examples include 1- or 2-assembly legal-weight transport casks described by Greene et al. (2013). The subcritical limit for fuel assemblies positioned in a tight array is on the order of four or fewer PWR assemblies (or BWR equivalent) by analogy to analysis of partial flooding of a horizontal, 32-PWR canister (EPRI 2008a, Section 3.4). Canisters could be de-rated by loading fewer than four PWR assemblies, if necessary to control criticality when flooded. Alternatively, subcriticality might be ensured by situating the repository in a host medium with saline groundwater (and not necessarily saturated brine as discussed for Design Alternative III.C). Criticality control without neutron absorber materials could have the advantage that the safety strategy would not rely on long lifetimes of those materials. Spent fuel loading and packaging steps could be performed in a fuel pool, or dry in a hot cell. Dewatering procedures might be modified (e.g., by operating with the canister in a heated cell; SKB 2010b, Section 6.1) if the small canisters do not get hot enough.

Example I.A.1 – Small capacity waste package, welded closure - *Canister closures could be welded without any need for further modification, for disposal in a wide range of geologic settings, if subcriticality is maintained when flooded with fresh (dilute) or mildly saline water (e.g., seawater).*

Example I.A.2 – Small capacity waste package, bolted closure – *Bolted canisters could maximize flexibility to modify the internals (e.g., to add fillers as in Design Alternative IV.B), or to unload the SNF and reuse the canister.*

Design Alternative I.B – Small capacity waste package combined with heavy insert.

Capacity: Up to approximately 4 PWR assemblies (or BWR equivalent) depending on enrichment and burnup.

Overpack: Mechanically robust for transport and handling. Provides containment for waste isolation, for a duration that is consistent with the safety strategy.

Canister and Basket: Maintains positions of assemblies after flooding with groundwater (to be consistent with analyzed configurations) for at least 10,000 years after waste package initial breach (which might occur soon after repository closure). Maintains material compatibility with a range of potential overpack designs.

Discussion: A few assemblies separated from each other within a solid insert also may require no absorbers when flooded with groundwater. Examples include the 4-PWR and 12-BWR SNF waste packages proposed for disposal in Sweden (SKB 2011). In the KBS-3 concept the cast iron insert is assigned to bear externally applied loads such as would be applied by the swelling clay buffer. It also reduces reactivity by spreading the assemblies apart, displacing moderator between them. Note that whereas a massive, monolithic cast iron insert could be appropriate for SNF disposal, other approaches to insert design and fabrication could be more appropriate for other waste forms (SKB 2013).

The SNF canister capacity could be limited by weight, since the KBS-3 insert alone weighs between 13.7 and 16.4 MT, compared to the total waste package weight of 24.6 to 26.8 MT (SKB 2010a, Section 1.7). For larger capacity (e.g., more than four PWR assemblies) some efficiency in total weight might result if criticality control could be achieved without further expanding the spacing between assemblies.

In the Swedish example, center-to-center spacings between assemblies are maintained (SKB 2011, Section 5.4.3; 37-cm separations, compared to 21-cm side dimensions for PWR assemblies). Burnup credit analysis is used for flooded canisters (SKB 2011, Section 8.4.1; $k_{eff} < 0.95$ including uncertainties). A burnup-enrichment loading curve was calculated for a reference waste package design (SKB 2010a, Table 2-3). Exceedance of this loading curve by individual assemblies may be possible based on package-specific analysis (SKB 2011, Section 5.3.4). Cast iron composition is controlled to limit neutron reflection (C<4.5% and Si<6%), and to limit gamma-embrittlement (Cu<0.05%)(SKB 2011, Section 5.4.1). Spent fuel loading and packaging steps could be performed in a fuel pool, or dry as proposed for the encapsulation plant for the Swedish repository (SKB 2011, Section 5.4.2).

One advantage of this alternative could be that degradation of the heavy cast iron insert will produce voluminous corrosion products that stabilize the fuel geometry and displace groundwater (compared to canisters with open spaces between assemblies). Iron corrosion products that form in reducing chemical environments, such as magnetite (Fe_3O_4), tend to form cohesive, compact layers (Smart 2008.) Whereas the KBS-3 concept uses copper as a corrosion-resistant outer layer on the waste package, for oxidizing environments a different corrosion resistant material could be used, and it could help maintain a reducing environment within the canister. Progressive degradation of the insert could increase neutron reflection but also increase moderator displacement by partly displacing interstitial water from within assemblies.

Example I.B.1 – Small capacity waste package combined with heavy insert, welded closure – *Like Example I.A.1, canister closures could be welded without any need for further modification, for disposal in a wide range of geologic settings, if subcriticality is maintained when flooded with fresh (dilute) or mildly saline water (e.g., seawater).*

4.2. Group II – Larger Canisters with Bolted Closures or Bolted/Welded Combinations

This group includes alternatives with larger capacity canisters (greater than 4-PWR size, possibly 9- or 12-PWR size or larger; or BWR equivalent) based on new designs. Bolted canisters could be reopened and resealed without disposing of major canister hardware. Heat output would be related to canister capacity and SNF characteristics, and would be independent of the type of closure. For canisters up to the size of existing DPCs, heat output and its effects would fall into the range of canisters already analyzed (Section 2.4).

For typical SNF enrichment and burnup, criticality control features would be needed for transportation and possibly disposal. Bolted closures could provide flexibility to match canister fuel loading and criticality control features, to future transportation and disposal requirements. Canisters with bolted closures could be initially loaded with one absorber configuration, then reopened later for modification as needed for disposal. Criticality control features to be added upon reopening (see Group IV) could be selected to serve both for transportation and disposal, once the disposal requirements are known (e.g., canisters loaded with PWR SNF in borated fuel pools). Alternatively, certain fuel assemblies could be removed prior to final packaging for disposal. Because the objective for Group II is flexibility at relatively low cost, basket structure and criticality control features built into the canisters at manufacture would likely not include long-lived materials such as those used for later Group V, but could be added later.

Design Alternative II.A – Bolted closures (both inner and outer canister lids).

Capacity: Any capacity up to limits imposed by handling, hoisting, and transportation.

Overpack: Mechanically robust for transport and handling. Provides containment for waste isolation, for a duration that is determined by the safety strategy.

Canister and Basket: Maintains positions of assemblies after flooding with groundwater (to be consistent with analyzed configurations) for at least 10,000 years after waste package initial breach (which might occur soon after repository closure). Maintains material compatibility with a range of potential overpack designs.

Discussion: Canisters with two bolted closures would offer maximum flexibility, but mechanical seals could be more costly to maintain (periodic testing requirements are typically required for bolted, but not welded closures).

Example II.A.1 – Larger canisters, bolted inner and outer closures – *One example of double-bolted closures is the CASTOR-V and CASTOR-X casks used at the Surry NPP in southeastern Virginia (Greene et al. 2013) and similar casks at many dry storage locations internationally, have two bolted lids. These are heavy, self-shielded storage casks licensed for storage but not transport in the U.S. Thin-walled canisters (with separate storage and transportation casks) with bolted lids are not generally available.*

Design Alternative II.B – Bolted closure combined with welded closure.

Capacity: Any capacity up to limits imposed by handling, hoisting, and transportation.

Overpack: Mechanically robust for transport and handling. Provides containment for waste isolation, for a duration that is determined by the safety strategy.

Canister and Basket: Maintains positions of assemblies after flooding with groundwater (to be consistent with analyzed configurations) for at least 10,000 years after waste package initial breach (which might occur soon after repository closure). Maintains material compatibility with a range of potential overpack designs.

Discussion: Same internal configurations possible as for bolted lids (Design Alternative II.A) but with one welded inner or outer lid. Possible advantages of combined welded and bolted closures include simplified containment monitoring during storage, and simplified accident analysis for transportation licensing.

Example II.B.1 – Larger canisters, bolted inner and welded outer closures - *The outer lid could be cut off and removed (e.g., dry) for access to the inner bolted lid without exposing fuel. The inner lid and shield could be readily removed for modification of internal components prior to disposal, then replaced and resealed. Canister design could add the facility to re-weld another outer lid (e.g., an extra chamfer).*

Example II.B.2 – Larger canisters, welded inner and bolted outer closures – *Removing the outer bolted lid would be relatively straightforward and could be done dry, but cutting out the inner lid would involve essentially the same choices as for existing (welded) DPC designs discussed in Group III.*

Design Alternative II.C – Can-within-can arrangements using bolted or welded closures.

Capacity: The inner canisters would be limited to 4-PWR capacity (or BWR equivalent), while the outer canisters could hold any number of inner canisters, up to limits imposed by handling, hoisting, and transportation.

Overpack: Mechanically robust for transport and handling. Provides containment for waste isolation, for a duration that is determined by the safety strategy. Disposal overpacks would contain the outer canister, which in turn would contain two or more inner canisters.

Canister and Basket: For the inner canisters: maintains positions of assemblies (consistent with analyzed basis), after flooding with groundwater, for at least 10,000 years after waste package initial breach (e.g., breach from early failure).

For the outer canister, maintains positions of inner canisters (consistent with analyzed basis), after flooding with groundwater, for at least 10,000 years after waste package initial breach. Maintains material compatibility with inner canisters, and with a range of potential overpack designs.

Discussion: The can-within-can ("canister-within-canister" of Scaglione et al. 2012) alternatives would package SNF in small canisters (up to 4-PWR size, or BWR equivalent), and provide a larger canister to hold two or more small ones. Sealing SNF in small canisters would preserve flexibility to use small waste packages (Group I) in the repository without exposing SNF during handling. The small inner canisters could be sealed in small overpacks for disposal using both enclosed and open emplacement modes. This approach would help to meet thermal and criticality control requirements for a wide range of disposal concepts, while also providing for handling, storage, and possibly transportation and disposal in a larger container.

Following the methods used currently for DPCs, shielding would be installed on the inner canisters, integral to the lid and/or as a separate shield plug. Shielding for the outer canister would be limited to that needed to control radiation scattered from the inner canisters.

Differences between welded and bolted outer canister closures are somewhat indistinct, so these options are not identified below as separate examples. Inner canisters, regardless of type, would tend to limit contamination of the outer canister which could potentially be reused or disposed of in a landfill. Bolted closure of the outer canister, or design features to allow welding a replacement outer lid after reopening, might be preferred to allow later installation of a filler (Group IV) in the interstitial volume between the inner canisters. Such capability could require additional ports or pass-throughs as part of the original outer canister basket design.

Loading of the smaller and more numerous inner canisters could impose operational burdens on fuel handling facilities (e.g., at NPPs), especially the time required for canister sealing and dewatering.

Example II.C.1 – Larger can-within-can systems with welded inner/bolted or welded outer canisters – *Inner canisters would have welded closures (similar to Example I.A.1), and would be cool and nonreactive enough to never require reopening prior to disposal. The outer canister lid could be welded or bolted. The inner canisters could be credited for storage and transportation licensing as containment envelopes, instead of, or in addition to the outer canister.*

Example II.C.2 – Larger can-within-can systems with bolted inner/bolted or welded outer canisters – *Inner canisters would have bolted closures (similar to Example I.A.2). These would also be cool and non-reactive enough to never require reopening prior to disposal, and bolted closures could be faster to implement. The outer canister lid could be welded or bolted, but a welded lid could simplify monitoring of canister integrity during storage.*

4.3. Group III – Existing DPC Designs (Welded Closures) Used for Disposal without Reopening

For this group of alternatives, neutron absorber materials such as Boral® which are used for criticality control during storage and transportation, are assumed to corrode in fewer than 10,000 years if the canister is breached. Waste isolation performance would be allocated to the disposal overpack, and thermal management would be described by the salt concept or an open emplacement mode (Section 2.4).

Group III consists of existing DPC designs without internal modifications. Heat output would be related to canister capacity and SNF characteristics. For canisters up to the size of existing DPCs, heat output and its effects would fall into the range of canisters already analyzed (Section 2.4). Postclosure criticality control would be assigned to external features (overpack, other barriers) or to non-reactive fuel characteristics. Group III is unique in assigning control to external features, because it represents possible pathways for direct disposal of existing DPCs without reopening.

Design Alternative III.A – High reliability (10^{-8}) overpack performance.

Capacity: Consistent with existing DPC designs and loading practices.

Overpack: Mechanically robust for transport and handling. Provides high-reliability containment in the disposal environment. Maintains material compatibility to limit corrosion or mechanical processes that could promote degradation of the fuel canister or basket.

Canister and Basket: Maintains configuration and neutron absorption for at least 10,000 years (without overpack breach, or until breach by disruptive events).

Discussion: This alternative would prevent criticality by preventing flooding of the waste canister for 10,000 years or longer, relying mainly on the disposal overpack. The probability of breach would need to be less than approximately 10^{-8} per package, from all causes including manufacturing defects. Overpack failure would be defined as degradation to an extent that allows flooding of the fuel canister and basket, and would therefore include localized corrosive attack. The overpack could be designed with thickness and materials such that corrosion slows with time. Note that reliance on unsaturated conditions with redundant engineered and natural barriers, is described by III.B below.

Overpacks would be designed, fabricated, loaded, and emplaced so that integrating over 10^4 overpacks gives a total probability less than 10^{-4} (per repository realization)of breach (and flooding with relatively rapid degradation of the neutron absorber components and possibly the basket). This alternative could require that the probability of manufacturing defects is small, i.e., less than previous estimates (DOE 2008). Also, at such low probability levels other hazards such as seismic ground motion and faulting could be important even for geologic settings that are not normally associated with these hazards.

This alternative could be challenging because of the high overpack reliability and quiescent geological conditions that could be needed to make it work. However, it might be part of a low-consequence argument whereby the total probability of criticality is greater than 10^{-4} per repository realization but still small, and the risks are shown by additional analysis to be insignificant.

Example III.A.1 – Existing DPC designs with high-reliability overpacks – *High-reliability disposal overpack performance prevents breach and DPC flooding and degradation, in a saturated or unsaturated environment, for at least 10,000 years.*

Design Alternative III.B – Disposal in unsaturated conditions with multiple engineered and natural groundwater exclusion/diversion barriers.

Capacity: Consistent with existing DPC designs and loading practices.

Overpack: Mechanically robust for transport and handling. Provides high-reliability containment in the disposal environment. Maintains material compatibility to limit corrosion or mechanical processes that could promote degradation of the fuel canister or basket.

Canister and Basket: Maintains configuration and neutron absorption for at least 10,000 years (without overpack breach, or until breach by disruptive events). Degrade slowly under disposal conditions, in a humidity environment, after waste package breach (consistent with analyzed basis).

Discussion: This alternative precludes criticality by effectively preventing waste package flooding, relying on unsaturated natural conditions (especially low groundwater flux), and redundant engineered barriers. This alternative was previously analyzed by EPRI (2008b).

Sites exist where groundwater is too scarce to flood a waste package, such as the deep unsaturated alluvium of Yucca Flat, Nevada (zero net infiltration except during glacial pluvial periods; Hardin et al. 2012a, Appendix B). Other unsaturated sites may have greater percolation flux, but redundant natural and engineered diversion and exclusion barriers could significantly reduce the probability of canister flooding.

The disposal environment would likely be humid even if groundwater intrusion did not occur, and degradation of SNF in a humidity environment can produce secondary hydrous precipitates. The reactivity of the SNF waste form altered to a form such as metaschoepite (a low-temperature hydrated secondary uranium mineral) would need to be evaluated, although previous analyses have excluded this as a process leading to criticality (BSC 2008e, Section 6.2.5). For unsaturated conditions (not flooded) there would be substantial reactivity margin to accommodate degradation of neutron absorbers.

In this alternative, the probability of flooding a breached canister would be shown to be insignificant because of the performance of the overpack and other barriers outside the canister. For two engineered barriers, a sufficiently low probability of flooding might be achieved for 10^4 packages, if the barriers behave independently, the individual barrier failure rate is small, and the incidence of water as seepage is limited. Independent barrier performance would be demonstrated given the possibility of possible common initiating events such as seismicity or climate change. To be successfully implemented, this alternative could require some combination of high-reliability overpacks (Design Alternative III.A), low probability of seepage into breached packages, and slow degradation of canister internals and SNF, to achieve a total probability of criticality less than 10^{-4} in 10,000 years, per repository realization.

Example III.B.1 – Existing DPC designs protected by multiple barriers – *Multiple natural and redundant engineered barriers prevent DPC flooding in an unsaturated disposal environment, for at least 10,000 years.*

Design Alternative III.C – Flooding possible only with chloride brine.

Capacity: Consistent with existing DPC designs and loading practices.

Overpack: Mechanically robust for transport and handling. Provides limited containment as needed in the salt disposal environment. Reacts with brine to form hydrogen gas and solid precipitates, consuming moisture that migrates toward waste packages. Maintains material compatibility to limit corrosion or mechanical processes that could promote degradation of the fuel canister or basket.

Canister and Basket: Maintains positions of assemblies (without overpack breach, or after breach for disrupted scenarios, consistent with analyzed basis).

Discussion: The possibility of waste package flooding in a salt repository is remote because moisture is scarce after repository closure. Intact and crushed salt in the near-field environment would be heated and dried out, then reconsolidated under lithostatic load. After a few hundred years when thermal conditions return to near-ambient, and rock stress conditions return to lithostatic, potential gradients for brine movement would be greatly reduced. Limited moisture that migrated toward waste packages (e.g., from potential gradients associated with gradients in brine saturation) would react with the disposal overpack. The overpack would have sufficient thickness of reactive, iron-bearing, corrosion-allowance material (i.e., steel) to react with such brine without resulting in breach.

Criticality analysis of existing, as-loaded DPCs (Scaglione and Clarity 2013) suggests that flooding with NaCl solutions could decrease the neutron multiplication factor (k_{eff}) on the order of -25%. This results because natural chlorine is 75% Cl-35, an effective neutron absorber. This calculated reduction in reactivity occurs even for loss-of-absorber and basket degradation cases. Package flooding in a salt repository might only be possible in a stylized scenario of human-caused intrusion from drilling of oil-and-gas wells. The drilling fluid in such wells is nearly always saline, and is often a chloride brine. Such drilling fluid, or other fluids produced in such wells, could be shown to have sufficient chloride concentration to prevent criticality. Seawater (~0.5 molal NaCl) is typical of saline groundwaters in many (non-evaporite) geologic settings, and the k_{eff} decrease for seawater would be only a few percent. Accordingly, this alternative would likely be limited to repositories within evaporite deposits, and particularly salt repositories.

Example III.C.1 – Existing DPC designs in a salt repository – *Disposal of DPCs in a salt repository, with thick-walled steel overpacks, prevents flooding and mitigates the consequences of any condition that might cause flooding (i.e., with chloride brine).*

Design Alternative III.D – Disposal of existing DPCs containing relatively non-reactive SNF, so that packages are subcritical after flooding and degradation of canister internals.

Capacity: Consistent with existing DPC designs and loading practices.

Overpack: Mechanically robust for transport and handling. Provides high-reliability containment in the disposal environment. Maintains material compatibility to limit corrosion or mechanical processes that could promote degradation of the fuel canister or basket.

Canister and Basket: Maintains configuration and neutron absorption until waste package breach. Degrades slowly under disposal conditions after waste package breach (consistent with analyzed basis).

Discussion: Criticality is prevented because the SNF is relatively non-reactive and remains subcritical even when canisters are flooded, and neutron absorbers and basket structure are degraded. The possibility of relatively non-reactive SNF was evaluated by Clarity and Scaglione (2013) for selected, actual DPCs. They found that some, but not all existing DPCs could be subcritical if flooded with pure water, for the degradation cases analyzed (see Section 2.5).

For implementation this alternative would benefit from high-reliability overpack performance, and slow overpack degradation for delayed breach. The probability of criticality will be related to: 1) uncertain degradation rates for the overpack and canister internals; 2) the uncertain conceptual bases for degraded fuel configurations; 3) the probabilities for different types of misloads; and 4) the uncertainty inherent to criticality analysis.

Example III.D.1 – Existing DPC designs loaded with relatively non-reactive SNF – *Canisters remain subcritical even after flooding with groundwater of dilute composition, and degradation of basket structure and neutron absorbers. Demonstrated using analysis of degraded configurations, with as-loaded assembly arrangements and fuel burnup characteristics.*

4.4. Group IV – Existing DPC Designs (Welded Closures) Used for Disposal with Modifications

This group of alternatives considers how existing DPC designs could be modified at fuel loading, or in the future after canister reopening, to mitigate the potential for postclosure criticality. Reopening could be limited to accessing the dewatering ports, or it could include cutting off lids. The canister would be reused (re-packaging is beyond the scope of this report). The basket and neutron absorber materials used in existing DPC designs are assumed to corrode readily if a canister is flooded with groundwater. For similar approaches for canisters with bolted closures, see Group II.

Like Group II, which is also based on existing DPC designs, heat output would be related to canister capacity and SNF characteristics, and would be independent of internal modifications. For canisters up to the size of existing DPCs, heat output and its effects *outside* the waste package would fall into the range of canisters already analyzed (Section 2.4). Thermal effects *inside* the canisters could be strongly impacted by modification (e.g., fillers, see Design Alternative IV.B).

Design Alternative IV.A – Reactivity control improvements.

Capacity: Consistent with existing DPC designs, but may be reduced as one possible measure to control reactivity.

Overpack: Mechanically robust for transport and handling. Provides containment for waste isolation, for a duration that is determined by the safety strategy. Maintains material compatibility to limit corrosion or mechanical processes that could promote degradation of the fuel canister or basket.

Canister and Basket: Maintains configuration and neutron absorption for at least 10,000 years (without overpack breach). Degrades slowly under disposal conditions (consistent with analyzed basis) after waste package breach.

Discussion: Reactivity control improvements could include fuel rearrangement, de-rating canisters by loading fewer assemblies, and installation of control rods. They were proposed by EPRI (2008b) for MPC-32 canisters, using "surrogate" control rod assemblies. Surrogate control rods would replace, or be added to, reactor control rods in some or all assemblies. Any irradiated control rods or other hardware that were removed would be managed as a separate waste stream (e.g., as "greater-than-Class-C" waste defined in regulation 10CFR Part 72).

Control rods could be designed to degrade at the same rate and in a similar manner to irradiated fuel rods, in any disposal environment, so that the geometrical association of neutron absorbing material with fissile fuel material in the waste package, would not be significantly changed. Importantly, this alternative could require a new type of criticality analysis that represents configurations during progressive degradation of fuel and control components, and of the basket structure.

This alternative would modify initial loading of existing DPC designs, either prior to initial closure welding, or after cutting open the canisters and rewelding. Flexibility to reopen canisters later and adjust to disposal requirements could be maximized by using the same reactivity improvement measures in bolted canisters (see Design Alternatives II.A and II.B). This alternative is a possible transitional strategy between the use of existing canisters without modification (Group III) and the use of new designs (Group V).

Example IV.A.1 – Reactivity control improvements in DPCs at initial loading – *Future disposal of existing DPC designs, with reactivity control improvements installed at initial fuel loading, for postclosure criticality control.*

Example IV.A.2 – Reactivity control improvements in reopened/rewelded DPCs – *Future disposal of existing DPC designs, with reactivity control improvements installed after cutting canisters open. Canisters would then be modified and rewelded.*

Design Alternative IV.B – Fillers for reactivity control.

Capacity: Consistent with existing DPC designs and loading practices.

Overpack: Mechanically robust for transport and handling. Provides containment for waste isolation, for a duration that is determined by the safety strategy. Maintains material compatibility to limit corrosion or mechanical processes that could promote degradation of the fuel canister or basket.

Canister and Basket: Maintains configuration and neutron absorption until waste package breach. Facilitates filler installation with dewatering ports and pass-throughs in the basket ("mouse holes"). Degrades slowly under disposal conditions after waste package breach, interacting with filler material (consistent with analyzed basis).

Discussion: Fillers would displace groundwater from the interstitial volume of fuel assemblies, and could also include neutron absorbing components. Fillers would be designed to remain immobile (e.g., solid) for at least 10,000 years after package breach. They would be resistant to temperatures reached immediately after installation, and facilitate heat transfer to limit SNF cladding temperature to 350°C or lower (NRC 2003). They would maintain material compatibility with the basket, neutron absorber components, and fuel.

Filler material could be installed at initial loading or later upon canister reopening. Various filler delivery strategies are available, such as pumping or pouring in a chemically curing liquid mixture or foam, or emplacing particulate materials that react with groundwater. Filler installation could be done dry in a hot cell, or possibly in a pool. The development approach would focus on candidate materials and their long-term performance. Investigation of filler materials and methods of installation is currently underway in the UFD R&D program. Previous work looked at the use fillers to mechanically stabilize SNF for transportation after extended storage (Maheras et al. 2012).

Like other reactivity control improvements (Design Alternative IV.A) this is a possible transitional strategy between the use of existing canisters without modification (Group III) and the use of new designs (Group V).

Example IV.B.1 – Fillers installed in DPCs at initial loading – *Future disposal of existing DPC designs, with filler installed at initial loading, for postclosure criticality control.*

Example IV.B.2 – Fillers installed in reopened/rewelded DPCs – *Future disposal of existing DPC designs, with filler installed after reopening (e.g., via dewatering ports). If canisters are cut open for filler installation (i.e., lids removed), the canister might be reused, for example by rewelding lids or by welding on a new head that fits on the canister end.*

4.5. Group V – Multi-Purpose Canisters with Long-Lived Baskets and Neutron Absorbers

This group includes alternatives with larger capacity canisters (e.g., 12-PWR size or larger) based on new designs, which would require criticality control features for transportation and possibly disposal, designed for lifetime of at least 10,000 years. Basket structure and criticality control components would be designed to maintain initial configuration, and preserve required neutron absorptivity, for at least 10,000 years after waste package breach (thus addressing the possibility for breach soon after repository closure). This group of alternatives is similar to Group II (larger canisters with bolted closures) but with: 1) long-lived internals; 2) either bolted or welded closures; and 3) criticality control through basket design rather than sole reliance on overpack reliability or added reactivity control improvements. This group of alternatives would attempt to envelope disposal requirements, with a canister design that meets performance requirements in a range of possible disposal environments. In the following discussion the range of disposal environments is subdivided to distinguish a design solution developed previously for oxidizing conditions.

Design Alternative V.A – Larger canisters, with basket and absorber materials selected for longevity in oxidizing environments.

Capacity: Any capacity up to limits imposed by handling, hoisting, and transportation.

Overpack: Mechanically robust for transport and handling. Provides containment for waste isolation, for a duration that is determined by the safety strategy. Maintains material compatibility to limit corrosion or mechanical processes that could promote degradation of the fuel canister or basket.

Canister and Basket: Maintains configuration and neutron absorption for at least 10,000 years (without overpack breach). Degrades slowly under disposal conditions (consistent with analyzed basis) after waste package breach. Maintains material compatibility with a range of potential overpack designs.

Discussion: This approach was adopted for postclosure criticality control, as part of a previous design for disposal of commercial SNF using the TAD canister (DOE 2008; 2006). The canister and basket structure would be fabricated from Type 316 (nuclear grade) stainless steel, and the neutron absorber plates from a stainless steel-boron alloy. Thicknesses would be prescribed based on measured corrosion rate data for appropriate environments, for corrosion allowance, to ensure subcriticality for at least

10,000 years after waste package breach. This could control criticality in case of early waste package failure from any cause.

Example V.A.1 – Larger canisters/baskets designed for ≥10,000-year life exposed to oxidizing disposal environments – *Larger canisters with stainless steel based canister shell and basket materials, designed to maintain initial SNF basket configuration and neutron absorption for >10,000 years exposure to groundwater or humidity environments, in oxidizing conditions.*

Design Alternative V.B – Larger canisters, with basket and absorber materials selected for longevity in a range of chemically reducing and oxidizing environments.

Capacity: Any capacity up to limits imposed by handling, hoisting, and transportation.

Overpack: Mechanically robust for transport and handling. Provides containment for waste isolation, for a duration that is determined by the safety strategy. Maintains material compatibility to limit corrosion or mechanical processes that could promote degradation of the fuel canister or basket.

Canister and Basket: Maintains configuration and neutron absorption for at least 10,000 years (without overpack breach). Degrades slowly under disposal conditions (consistent with analyzed basis) after waste package breach. Maintains material compatibility with a range of potential overpack designs.

Discussion: This alternative could implement a similar configuration to that for oxidizing conditions (Design Alternative V.A) but with structural and neutron absorbing components fabricated from materials with corrosion resistance over a broader range of environmental conditions. Chemically reducing conditions, while limiting the mobility of most waste radionuclides in aqueous transport, can require different engineered materials to achieve corrosion resistance, than oxidizing conditions. The possibility of radiolytic oxidation complicates the chemical environment, and could add to the advantages from using materials that remain corrosion resistant over a broad range of conditions. Materials such as Alloy 22 could exhibit such corrosion resistance (Rebak and Crook 2004) and might be used for basket structures and neutron absorber components. Other materials such as copper and titanium have also been identified (Rebak 2007). Application of such materials in canister and basket designs is developmental. International programs have generally not investigated corrosion resistant neutron absorbing materials because they have pursued small canisters (Group I).

Example V.B.1 – Larger canisters/baskets designed for ≥10,000-year life exposed to reducing or oxidizing disposal environments – *Larger canisters with corrosion resistant canister shell and basket materials, designed to maintain initial SNF basket configuration and neutron absorption for >10,000 years exposure to groundwater or humidity environments, in reducing or oxidizing conditions.*

CONCLUSION

Many different alternatives are available for SNF canisters that would be used for storage, transportation, and disposal (MPCs). A total of 13 Design Alternatives with 19 examples are identified and organized in five groups in Table 1 and Section 4. Current DPC designs reflect only Group III, although some earlier designs with bolted closures reflect Group II.

The applicability of Design Alternatives to different disposal concepts, including enclosed and open emplacement modes (Hardin et al. 2013a) and different types of geologic host media, is shown in Table 2. The small canisters (Group I) could support disposal in any geologic setting. The bolted canisters (Group II) including the can-within-can alternative (Design Alternative II.C) appear to have excellent flexibility to support implementation of various disposal concepts in different geologic settings. The salt disposal concept and the open modes can accept larger canisters containing more SNF assemblies. Existing DPCs may be directly disposable without modification (Group III) but only from canisters (as-loaded) and for some disposal concepts and geologic settings. The range of possible disposal solutions for existing DPCs is expanded if they can be reopened and modified (Group IV). It may be possible to develop an enveloping MPC design (Group V) that is directly disposable with a range of disposal concepts and geologic settings.

Some of the Design Alternatives are incompatible with certain design concepts and geologic settings. For example, Groups II through V represent canisters that could be too large (or would contain too many SNF assemblies) for enclosed emplacement modes. Similarly, Alternatives III.B and V.A are specific to unsaturated settings, and Alternative III.C is specific to the salt concept. Smaller canisters may be disposable in any geologic setting, but some disposal concepts may be infeasible if disposal conditions permit the use of larger canisters holding more SNF.

Some of the alternatives could be transitional, allowing direct disposal of existing DPCs with or without modifications (Groups IV and III, respectively), while other solutions are being prepared. Bolted closures or combinations of bolted and welded closures (Groups I and II) offer flexibility to respond to future fuel management decisions (e.g., repository siting, or a decision to reprocess SNF) that require re-packaging or modification of canister internals.

Estimated cost savings from implementing MPCs in the U.S. nuclear waste management system for SNF dry storage, and later transportation and disposal, depend on when such measures are introduced. Timely implementation in the next 10 to 20 years is projected to save on the order of $10 billion in re-packaging costs (Hardin et al. 2013a, Section 9). The capability to dispose of SNF in larger packages (e.g., larger than 4-PWR size) would allow similar additional savings in disposal costs (Kalinina and Hardin 2012). Such savings could approach a third of the total cost of disposal (Hardin et al. 2012a, Section 5). The outlook for deploying MPCs depends on progress in disposal planning and repository siting, and cooperation of the nuclear power utilities, the vendor industry, and government.

REFERENCES

Andra (National Radioactive Waste Management Agency) 2005. *Dossier 2005 argile–architecture and management of a geological disposal system*. December, 2005. http://www.Andra.fr/ international/ download/Andra-international-en/document/editions/268va.pdf.

BSC (Bechtel SAIC Co.) 2007a. *General Corrosion and Localized Corrosion of Waste Package Outer Barrier*. U.S. Department of Energy, Office of Civilian Radioactive Waste Management. ANL-EBS-MD-000003 REV 03. July, 2007.

BSC (Bechtel SAIC Co.) 2007b. *Analysis of Mechanisms for Early Waste Package / Drip Shield Failure*. U.S. Department of Energy, Office of Civilian Radioactive Waste Management. ANL-EBS-MD-000076 Rev. 0. June, 2007.

BSC (Bechtel SAIC Co.) 2008a. *Postclosure Analysis of the Range of Design Thermal Loadings*. U.S. Department of Energy, Office of Civilian Radioactive Waste Management. ANL-NBS-HS-000057 Rev. 0. January, 2008.

BSC (Bechtel SAIC Co.) 2008c. *Features, Events, and Processes for the Total System Performance Assessment: Analyses*. U.S. Department of Energy, Office of Civilian Radioactive Waste Management. ANL-WIS-MD-000027 REV 0. March, 2008.

BSC (Bechtel SAIC Co.) 2008d. *Screening Analysis of Criticality Features, Events, and Processes for License Application*. U.S. Department of

Energy, Office of Civilian Radioactive Waste Management. ANL-DS0-NU-000001 Rev. 0. February, 2008.

BSC (Bechtel SAIC Co.) 2008e. *CSNF Loading Curve Sensitivity Analysis*. U.S. Department of Energy, Office of Civilian Radioactive Waste Management. ANL-EBS-NU-000010 Rev. 0. February, 2008.

Carter, J.T., F. Hansen, R. Kehrman and T. Hayes 2011. *A generic salt repository for disposal of waste from a spent nuclear fuel recycle facility*. SRNL-RP-2011-00149 Rev. 0. Aiken, SC: Savannah River National Laboratory.

Clarity, J.B. and J.M Scaglione 2013. *Feasibility of Direct Disposal of Dual-Purpose Canisters-Criticality Evaluations*. ORNL/LTR-2013/213. Oak Ridge National Laboratory, Oak Ridge, TN. June, 2013.

DOE (U.S. Department of Energy) 1994. *Multi-Purpose Canister System Evaluation*. DOE/RW-0445. Office of Civilian Radioactive Waste Management. Washington, D.C.

DOE (U.S. Department of Energy) 2003. *Disposal Criticality Analysis Methodology Topical Report.* YMP/TR-004Q Rev. 2. Office of Civilian Radioactive Waste Management. November, 2003.

DOE (U.S. Department of Energy) 2006. *Preliminary Transportation, Aging and Disposal Canister System Performance Specification.* DOE/RW-0585 Rev. A. Office of Civilian Radioactive Waste Management. November, 2006.

DOE (U.S. Department of Energy) 2008. *Yucca Mountain Repository License Application for Construction Authorization*. DOE/RW-0573. Washington, D.C.: U.S. Department of Energy.

EPRI (Electric Power Research Institute) 2008a. *Feasibility of Direct Disposal of Dual-Purpose Canisters: Options for Assuring Criticality Control.* #1016629. Palo Alto, CA.

EPRI (Electric Power Research Institute) 2008b. *Feasibility of Direct Disposal of Dual-Purpose Canisters in a High-Level Waste Repository*. #1018051. Palo Alto, CA.

Graf, R., K.-J. Brammer and W. Filbert 2012 (in German). "Direkte Endlagerung von Transport- und Lagerbehältern - ein umsetzbares technisches Konzept." *Jahrestagung Kerntechnik 2012*, Stuttgart, May, 2012.

Greene, S.R., J.S. Medford and S.A. Macy 2013. *Storage and Transport Cask Data for Used Commercial Nuclear Fuel – 2013 U.S. Edition*. ATI-TR-13047. Energx, Oak Ridge, TN, and Advanced Technology Insights, LLC, Knoxville, TN.

Hansen, F.D., E.L. Hardin, R.P. Rechard, G.A. Freeze, D.C. Sassani, P.V. Brady, C.M. Stone, M.J. Martinez, J.F. Holland, T. Dewers, K.N. Gaither, S.R. Sobolik, and R.T. Cygan 2010. *Shale Disposal of U.S. High-Level Radioactive Waste*. SAND2010-2843. Albuquerque, NM: Sandia National Laboratories. May, 2010.

Hansen, F.D. and M.K. Knowles 2000. "Design and analysis of a shaft seal system for the Waste Isolation Pilot Plant." *Reliability Engineering and System Safety* 69 (2000), pp. 87-98.

Hansen, F.D., and C.D. Leigh 2011. *Salt Disposal of Heat-Generating Nuclear Waste*. SAND2011-0161. OSTI ID: 1005078. Albuquerque, NM: Sandia National Laboratories. January, 2011.

Hardin, E. and D. Sassani 2011. "Application of the Prefabricated EBS Concept in Unsaturated, Oxidizing Host Media." Proceedings: 13th International High-Level Radioactive Waste Management Conference, Albuquerque, NM. April, 2011. American Nuclear Society. Paper #3380.

Hardin, E., J. Blank, H. Greenberg, M. Sutton, M. Fratoni, J. Carter, M. Dupont and R. Howard 2011. *Generic Repository Design Concepts and Thermal Analysis (FY11)*. FCRD-USED-2011-000143 Rev. 0. U.S. Department of Energy, Used Fuel Disposition Campaign.

Hardin, E., T. Hadgu, D. Clayton, R. Howard, H. Greenberg, J. Blink, M. Sharma, M. Sutton, J. Carter, M. Dupont and P. Rodwell 2012a. *Repository Reference Disposal Concepts and Thermal Load Management Analysis*. FCRD-UFD-2012-00219 Rev. 2. U.S. Department of Energy, Used Fuel Disposition Campaign.

Hardin, E., T. Hadgu, H. Greenberg and M. Dupont 2012b. *Parameter Uncertainty for Repository Thermal Analysis*. FCRD-UFD-2012-000097 Rev. 0. April, 2012. U.S. Department of Energy, Used Fuel Disposition R&D Campaign.

Hardin, E. and R. Howard 2013. *Assumptions for Evaluating Feasibility of Direct Geologic Disposal of Existing Dual-Purpose Canisters*. FCRD-USED-2012-000352 Rev. 1. November, 2013. U.S. Department of Energy, Used Fuel Disposition R&D Campaign.

Hardin, E. and M. Voegele 2013. *Alternative Concepts for Direct Disposal of Dual Purpose Canisters*. FCRD-UFD-2013-000102, Rev.0. U.S. Department of Energy, Used Fuel Disposition R&D Campaign.

Hardin, E.L., D.J. Clayton, R.L. Howard, J.M. Scaglione, E. Pierce, K. Banerjee, M.D. Voegele, H.R. Greenberg, J. Wen, T.A. Buscheck, J.T. Carter, T. Severynse and W.M. Nutt 2013a. *Preliminary Report on Dual-Purpose Canister Disposal Alternatives (FY13)*. FCRD-UFD-2012-

000170 Rev. 0. U.S. Department of Energy, Used Fuel Disposition R&D Campaign.

Hardin, E., C. Stockman, E. Kalinina and E.J. Bonano 2013b. "Integration of Long-Term Interim Storage with Disposal." *OECD/NEA International Workshop on Safety of Long Term Interim Storage Facilities*. May 21–23, 2013. Munich, Germany.

Hardin, E., D.J. Clayton, M.J. Martinez, G. Neider-Westermann, R.L. Howard, H.R. Greenberg, J.A. Blink and T.A. Buscheck 2013c. *Collaborative Report on Disposal Concepts*. FCRD-UFD-2012-000170 Rev. 0. U.S. Department of Energy, Used Fuel Disposition R&D Campaign.

Howard R., J.M. Scaglione, J.C. Wagner, E. Hardin and W.M. Nutt. 2012. *Implementation Plan for the Development and Licensing of Standardized Transportation, Aging, and Disposal Canisters and the Feasibility of Direct disposal of Dual Purpose Canisters*. FCRD-UFD-2012-000106 Rev. 0. U.S. Department of Energy Fuel Cycle Technology Program, Used Fuel Disposition Campaign.

Maheras, S.J., R.E. Best, S.B. Ross, E.A. Lahti and D.J. Richmond 2012. *A Preliminary Evaluation of Using Fill Materials to Stabilize Used Nuclear Fuel During Storage and Transportation.* FCRD-UFD-2012-000243. August, 2012. U.S. Department of Energy, Used Fuel Disposition R&D Campaign.

Mariner, P.E., J.H. Lee, E.L. Hardin, F.D. Hansen, G.A. Freeze, A.S. Lord, B. Goldstein and R.H. Price 2011. *Granite disposal of U.S. high-level radioactive waste*. SAND2011-6203. Albuquerque, NM: Sandia National Laboratories.

NAGRA (Swiss National Cooperative for the Disposal of Radioactive Wastes) 2002. *Project Opalinus clay safety report*. NAGRA Technical Report 02-05. December, 2002.

NAGRA (Swiss National Cooperative for the Disposal of Radioactive Wastes) 2003. *Canister options for the disposal of spent fuel*. NAGRA Technical Report NTB 02-11.

NAGRA (Swiss National Cooperative for the Disposal of Radioactive Wastes) 2009. *A review of materials and corrosion issues regarding canisters for disposal of spent fuel and high-level waste in Opalinus Clay*. NAGRA Technical Report NTB 09-02.

NRC (U.S. Nuclear Regulatory Commission) 2003. *Cladding Considerations for the Transportation and Storage of Spent Fuel*. Interim Staff Guidance SFST-ISG-11, Rev. 3. November 17, 2003.

NRC (U.S. Nuclear Regulatory Commission) 2013. *Waste Confidence Generic Environmental Impact Statement – Draft Report for Comment.* NUREG-2157. Office of Nuclear Material Safety and Safeguards. ML13224A106. (http://www.nrc.gov/waste/spent-fuel-storage/wcd/ documents.html#tech)

Rebak, R.B. and Crook, P. 2004. "Influence of the Environment on the General Corrosion Rate of Alloy 22 (N06022)." *Proc. American Society of Mechanical Engineers Pressure Vessels and Piping*. July 25-29, 2004. San Diego, CA. (UCRL-PROC-203671).

Rebak, R.B. 2007. "Environmental Degradation of Materials for Nuclear Waste Repositories Engineered Barriers." *Proc. 3rd International Environmental Degradation of Engineering Materials (EDEM 2007)*. May 21-25, 2007. Gdansk, Poland. (UCRL-PROC-227056).

Rigby, D.B. 2010. *Evaluation of the Technical Basis for Extended Dry Storage and Transportation of Used Nuclear Fuel*. U.S. Nuclear Waste Technical Review Board. December, 2010. (www.nwtrb.gov).

Scaglione, J.M. A.G. Caswell and G. Radulescu 2012. *Status Report on Integrated Canister Design and Evaluation.* ORNL/LTR-2012/448. Oak Ridge National Laboratory.

Shoesmith, D.W., B.M. Ikeda, M.G. Bailey, M.J. Quinn and D.M. LeNeveu 1995. *A Model for Predicting the Lifetimes of Grade-2 Titanium Nuclear Waste Containers*. AECL-10973. Atomic Energy-Canada Ltd., Whiteshell Laboratories, Pinawa, Manitoba.

SKB (Swedish Nuclear Fuel and Waste Management Co.) 2010a. *Fuel and canister process report for the safety assessment SR-Site.* Technical Report TR-10-46.

SKB (Swedish Nuclear Fuel and Waste Management Co.) 2010b. *Design, production and initial state of the canister.* Technical Report TR-10-14.

SKB (Swedish Nuclear Fuel and Waste Management Co.) 2011. *Long-term safety for the final repository for spent nuclear fuel at Forsmark: Main report of the SR-Site project, Volume I.* Technical Report TR-11-01.

SKB (Swedish Nuclear Fuel and Waste Management Co.) 2013. *Spent Fuel Geologic Consultation.* SKB International Report 166. Prepared for Savannah River Nuclear Solutions, LLC. Final Report, September, 2013.

Smart, N. 2008. "The Corrosion Behavior of Carbon Steel Radioactive Waste Packages: A Summary Review of Swedish and U.K. Research." National Association of Corrosion Engineers (NACE) – Corrosion 2008 Conference and Expo.

Stauffer, P.H., L.H. Auer and N.D. Rosenberg 1997. "Compressible gas in porous media: a finite amplitude analysis of natural convection." *Intl. J. Heat Mass Transfer*. Vol. 40, No. 7, pp. 1585-1589.

Wagner, J. and C. Parks 2001. *Recommendations on the Credit for Cooling Time in PWR Burnup Credit Analyses*. NUREG/CR-6781. U.S. Nuclear Regulatory Commission, Washington, D.C.

Wagner, J., C. Parks, D. Mueller and I. Gauld 2011. "NRC Research in Support of Burnup Credit Regulatory Guidance." U.S. Nuclear Regulatory Commission, 22nd Annual Regulatory Information Conference, March 9-11, 2011, North Bethesda, MD. (http://www.nrc.gov/public-involve/conference-symposia/ric/past/2010/slides/th32wagnerjhv.pdf).

In: Radioactive Waste
Editor: Susanna Fenton
ISBN: 978-1-63321-731-7

Chapter 3

NUCLEAR SECURITY SYSTEM IN GEORGIA

G. Nabakhtiani
Head of the First Division
Department for Nuclear and Radiation Safety
Georgian Ministry of Protection of Environment and Natural Resources
Tbilisi, Georgia

ABSTRACT

Georgia has received a difficult heritage from the former Soviet Union related to nuclear security situation. The country already took some steps to combat nuclear and radiation threats. Developing of nuclear forensics capability is one of the important steps to reach set goals to establish a nuclear security regime.

I. INTRODUCTION

Georgia is a small country situated on the territory of the south Caucasus region and neighbored by some countries having developed nuclear industries. The country had received difficult heritage from soviet period of time – a number of abandoned (orphaned) radioactive sources. There were found more than 300 such type sources in Georgia. Another big issue is export-import control: Georgia is a transit country; therefore many loads crossed the Georgian territory. Georgia receives support from international organizations and country to solve its problems related to nuclear security. The country became an IAEA member in 1996 and received serious assistance from the Agency on the different fields since 1997. Georgia under IAEA support developed special plans for implementation of the state nuclear security regime. The plan implementation can be focused on the following tasks:

- Elaborate legislative framework and define competent authorities with their responsibilities;
- Involvement in international legal instruments;
- Establishment of radioactive source inventory;
- Conduct regulatory and control activity (export-import and control inside the country) to prevent accidents.

In line with all above mentioned it is essential that country has capability conduct full scale nuclear forensic activity (Itself or supported internationally) to achieve the set goals for nuclear security.

II. Legislative Framework

According to the Code of Conducts: "Every State should have in place an effective national legislative and regulatory system of control over the management and protection of radioactive sources" [1]. The Georgian radiation safety system is based on the framework legislation – the Law of Georgia No.1674-IS "On Nuclear and Radiation Safety" – which put into the force on January 1, 1999. The new version of the law (No.5912) was adopted at 2012. This law (Article 6, Paragraph 1) designates the Ministry of Environment and Natural Resources Protection of Georgia (MENRP) as a national regulatory body in the field of nuclear and radioactive activities and with this purpose the Department for Nuclear and Radiation Safety (DNRS) is set up within the MENRP (Article 6, Paragraph 2). The control of any transfer of radioactive sources inside the country and export and import is regulated by the special system of licenses and permits, which is based on the framework legislation – the Law of Georgia No.1775-RS "On Licenses and Permits" – adopted on June 24, 2005. Every act of import and export of not exempted (or cleared) radioactive material must be approved by the special permit issued by the MENRP. Such a permit is based on the conclusion of the DNRS. Main regulatory norms are defined by national basic safety standard – RSL-2000.

(New version of national basic safety requirements in accordance with international norms is already drafted). Other laws and regulations (such as Law "On Transport of Radioactive Substances", regulations: "main rules for Transport of radioactive Substances", "On Security of Radioactive Sources" and others) are under process of elaboration. All existed and planned to put in force requirements provide strong basement for establishment of the country nuclear security regime considering all main essential elements [2]:

- State Responsibility;
- Identification and Definition of Nuclear Security Responsibilities;
- Legislative and Regulatory Framework;
- International Transport of Nuclear Material and Other Radioactive Material;
- Offenses and Penalties Including Criminalization;
- International Cooperation and Assistance;
- Identification and Assessment of Targets and Potential Consequences;
- Use of Risk-Informed Approaches;
- Detection of Nuclear Security Events;
- Planning for, Preparedness for, and Response to a Nuclear Security Event;
- Sustaining a Nuclear Security Regime

III. Involvement in International Legal Instruments

Georgia ratified the Convention on the Physical Protection of Nuclear Materials (CPPNM) at October 7, 2006. The amendments of the conventions – CPPNME also were adopted at April 05, 2012. Georgia also endorsed IAEA "Code of Conduct on Safety and Security of Radioactive Sources" .The country also signed up "Convention of the Suppression of the Act of Nuclear Terrorism" (has not ratified yet).

Georgia officially ratified "Convention on Early Notification of Nuclear Accident" and joined to "Treaty on the Non-Proliferation of Nuclear Weapons (NPT)" as non-nuclear weapon state.

The agreement between Georgia and IAEA for application of safeguards in connection with NPT entered into force on June 3, 2003 (INFCIRC/617) by

Law No. 2111. The additional protocol of the safeguards entered into force on the same day.

IV. Establishment of Radioactive Source Inventory

One important task for establishing the country's nuclear security regime is defining the area for the system application – establishment of inventory for radioactive sources and associated activities. Establishment of the inventory requires existing of system of information collection and processing, system of information maintenance and special computer code (software) for keeping and working with information.

Up to 2004 DNRS developed only partially completed inventory covered some main users of radiation sources. The information was kept as a hardcopy and in Excel files. The information was not comprehensive and sometimes not fully correct.

At the same time, information keeping did not meet strong security requirements [3]. According to IAEA Code of Conducts every state should have inventory of its radioactive sources. Receiving support from US NRC Georgian RA – DNRS had started activity to create full scale inventory of all sources of ionization radiation existed in Georgia. DNRSS was granted by computer code RASOD elaborated by Armenian specialists under US NRC programme to support of some former Soviet country in establishment of inventory of sources of ionization radiation.

Information collection was divided on the two main stages. At the first stage special letters were disseminated to the potential owners asking them to provide information for their radioactive sources. As a Georgian experience showed this type activity was not as effective as it was desired. Therefore special on-site checks were conducted. The checks had three specific goals:

1. Collect all information for the inventory;
2. Find out new information for other potential users;
3. Encourage the user to receive license on nuclear and radiation activity.

So, full scale inventory of ionization radiation sources (not only radioactive sources) and associated activities were created. The sources were grouped in classless according to IAEA requirements [4] what provides base to determine security level for different sources [3].

V. Regulatory and Control Activity

Every activity connected to not exempt sources of radiation should be authorized in Georgia. Every action of radioactive source owner changing or sources export-import requires issuing of special permit. These regulations provides good legislation base for strong security control for sources transfer.

As it was mentioned above Georgia has received a difficult heritage related to S.C. orphan radioactive sources. As a result several radiological accidents have been developed in Georgia since 1997. The first great radiological accident took place at the military base in Lilo, when 11 soldiers were irradiated by ^{137}Cs (orphan ^{60}Co and ^{226}Ra sources also were found) [5], when Georgia received support from the Agency within the project GEO/9/004.

To avoid occurrence of such incidents and accidents the special searching operations were conducted. There are three main possibilities to conduct searching operation: Airborne survey, car survey and pedestrian survey. The first possibility is the most effective and expensive at the same time. This type searching was carried out in Georgia within the scope of IAEA TC project GEO/9/006, when 56 hours of airborne gamma survey of a large territory of the western part of Georgia and around Tbilisi was carried out at 2000. During the operation one orphan radioactive source were found in Poti. Airborne survey is not effective for mountain relief. So, taking into account high price for this activity, car and pedestrian searching also were conducted at 2002, 2003 and 2005. All these activity were actively supported by the Agency in close collaboration with USA, France, Indian and Turkish experts.

The most important activity for combating of nuclear threats for this activity is nuclear forensics to identify the sources and possible pathways for losing of control, and possibility to use them for construction of S.C. "Dirty Bomb".

Among the found orphan sources the most important are RTG. Each of them contains radionuclide ^{90}Sr/^{90}Y (initial activity of ^{90}Sr is 1 290 TBq). Six RTG's were found and recovered. The sources were used to produce electric supply for antennas installed into the gorge of high mountain river Enguri [6]. Due to energy loss of radiation the sources are very hot, therefore using of thermocouples gives the possibility to receive enough electrical power to supply the antenna with energy. Usually the sources were installed into special device. Very often found orphan sources are military devices containing ^{137}Cs radionuclides. As it was mentioned above there were fixed two types of devices (special containers): The first contains one source with activity ~3Ci, the second two sources with activity ~10mCi for each. Figure 1 demonstrates distribution of found and recovered orphan source on radionuclides.

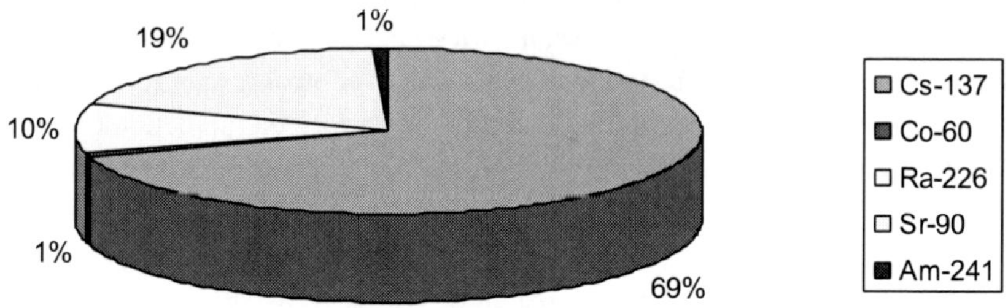

Figure 1. Found and recovered orphan radioactive sources[1].

To take into account main causes for loss of control over the sources, it is possible to conclude that the main aspect loss of control and originating of orphan sources was Financial Motive. This motive was existed when some people found abandoned radioactive sources. They just tried to earn money and improve their wealth in difficult economical situation. Based on above-mentioned there is possible to identify three main causes for originating of orphan radioactive sources in Georgia:

- Temporary absence of regulatory control;
- Absence of radioactive waste management system;
- Difficult economical situation

All found orphan sources were safely recovered. Before the starting of recovery activity all main actions developed according to ITWG Nuclear Forensics Plan was implemented [7], therefore in recovery operations official representatives of different organizations were involved: RA: Ministry of Interior Affairs (Special operation group, Emergency department), scientist laboratory. It should be noted that on field investigation always is conducted by RA specialists. It is possible to distinguish to type recovery operations:

[1]Figure 1 does not include RTG's.

recovery operation during searching activity and large scale recovery operation. Large scale recovery operation is required when powerful orphan radioactive source was found. Good example for this activity is recovery operation with RTG [6]. The conducted operation included three main phases: Gathering of information, situation assessment and implementation of planned activities.

During the first phase, nuclear forensic assistance is very important. During the second phase, the special trainings for rescuers should be conducted. Only in this way can be strictly define recovery team composition, responsibility of each team member and construction of equipment and container necessary for the recovery operation. This phase also considers feedback to regulation and planning of future activity to avoid such type accident repeating. The results of nuclear forensics is especially important for this phase to understand reasons why the incident (or accident occurred), what can be negative impact of the incident (accident)? This understanding will help to avoid occurrence of accident in future. As a result of conducted activities, the number of incidents connected to orphan radioactive sources was minimized.

Georgia is a transit country and usually many goods were crossed over the country territory, therefore it is most important task to provide security regime implementation at the country borders. This regime contains three main elements: legislative, technical and administrative. The legislative was just discussed briefly above.

To solve the technical issue the special portal monitors (based on the support of USA) were installed at the Georgia border check points. Especial legal requirement was adopted to define nuclear forensics activities at the border if suspicious radioactive material is found.

CONCLUSION

As a conclusion for all above mentioned can be said that nuclear forensics investigation is an important activity to establish nuclear security regime. Georgia has enough scientific potential to conduct forensic investigation. Simultaneously the country develops its capability to conduct nuclear forensics activity by the upgrading of technical means and administrative tools.

REFERENCES

[1] INTERNATIONAL ATOMIC ENERGY AGENCY, *Code of Conduct on Safety and Security of Radioactive sources*, Vienna (2004), p.5.

[2] INTERNATIONAL ATOMIC ENERGY AGENCY, Objectives and Essential Elements for State's Nuclear Security Regimes, IAEA Nuclear Security Series No.20, Nuclear Security Fundamentals, Vienna (2013).

[3] INTERNATIONAL ATOMIC ENERGY AGENCY, Security of Radioactive Sources, *IAEA Nuclear Security* Series No.11, Implementing Guide, Vienna (2009).

[4] INTERNATIONAL ATOMIC ENERGY AGENCY, Categorization of Radioactive Sources, *IAEA Safety Standards*, Safety Guide RS-G-1.9, Vienna (2005).

[5] INTERNATIONAL ATOMIC ENERGY AGENCY, *The Radiological Accident in Lilo,* Vienna (2000).

[6] INTERNATIONAL ATOMIC ENERGY AGENCY, *The Radiological Accident in Lia,* Georgia, Vienna (2013).

[7] INTERNATIONAL ATOMIC ENERGY AGENCY, Nuclear Forensics Support, *IAEA Nuclear Security* Series No.2, Technical Guidance, Vienna (2006).

INDEX

D

E

J

K

L

M

N

O

P

Q

R

S

T

U

V

W